Digital Control of Power Converters Using Arduino and an STM32 Microcontroller

This concise and accessible guide equips readers with the knowledge and skills needed to implement digital control algorithms to design efficient and reliable power converters using STM32 microcontrollers.

Through this book, Majid Pakdel covers a range of topics including digital control theory, switching converters theory, the design and implementation of control algorithms (such as proportional–integral–derivative and advanced digital control techniques), programming of STM32 microcontrollers, and interfacing with power electronics components. He also provides step-by-step tutorials and code examples to help readers understand and implement the concepts in their own projects. Readers will gain a deep understanding of digital control techniques in power converters, learn how to program STM32 microcontrollers for control applications, and be able to design and implement their own digital control algorithms in power electronics systems. The practical examples provided in the book will help readers apply the knowledge gained to real-world projects and improve their skills in developing digital control systems.

The information within is useful for young professionals and students aiming at experimental implementation on a microcontroller platform of a control algorithm for power converters. To fully benefit from the practical examples demonstrating digital controller implementation on the STM32, readers should have a solid understanding of power switching converter topologies, modeling, and control.

Majid Pakdel earned a bachelor's degree at Amirkabir University of Technology in 2004, a master's degree at Isfahan University of Technology in 2007, a PhD in electrical power engineering at the University of Zanjan in 2018, and a further master's degree in AI and robotics at Malek Ashtar University of Technology in 2023.

Digital Control of Power Converters Using Arduino and an STM32 Microcontroller

Majid Pakdel

CRC Press is an imprint of the
Taylor & Francis Group, an **informa** business

Designed cover image: Microcontrollers, PCB electrical boards, semiconductor microchip technology computers machine, circuit electronics concept hardware pi chip, neat vector illustration of electronic technology component; Shutterstock.

First edition published 2025
by CRC Press
2385 NW Executive Center Drive, Suite 320, Boca Raton FL 33431

and by CRC Press
4 Park Square, Milton Park, Abingdon, Oxon, OX14 4RN

CRC Press is an imprint of Taylor & Francis Group, LLC

ISBN: 978-1-032-89038-8 (hbk)
ISBN: 978-1-032-89137-8 (pbk)
ISBN: 978-1-003-54135-6 (ebk)

DOI: 10.1201/9781003541356

Typeset in Minion
by SPi Technologies India Pvt Ltd (Straive)

Contents

Preface

In recent years, power converters have become an integral part of modern electronics, providing efficient and reliable energy conversion for a wide range of applications. With the advancement of digital control technology, there is a growing interest in using microcontrollers such as Arduino and STM32 to implement control algorithms for power converters.

This book aims to provide a comprehensive guide to understanding and implementing digital control of power converters using Arduino and STM32 microcontrollers. It covers the fundamental principles of the design and implementation of control algorithms, and practical examples of real-world applications.

Whether you are an electrical engineering student, researcher, or practicing engineer, this book will help you develop the necessary skills to design and implement digital control systems for power converters. Through a combination of theoretical concepts and hands-on examples, you will gain a deep understanding of the principles and techniques required to successfully implement digital control systems in your own projects.

The author hopes this book will serve as a valuable resource for anyone interested in learning about digital control of power converters using Arduino and STM32 microcontrollers. Happy reading and happy experimenting!

A buck converter is a power electronic device whose job is to step down the input voltage to a desired output voltage to supply power to all sorts of electronic loads. The output voltage is sensed through $H(s)$ and compared to a desired output voltage to then generate an error signal. This error signal is then fed to our controller or $C(s)$ which generates and sends a continuous control voltage signal to the pulse-width modulation (PWM) driver driving the plant. The buck converter is duty cycle controller with the relation $V_o = D \times V_i$, where D varies between 0 and 1. The feedback controller $C(s)$ drives the PWM duty cycle value to keep the error at zero. However, let's not overlook how [i]t operates. The point we're trying to make is that this system is [e]ntirely analog controlled as there is no digital or discrete cou[p]ling between the blocks. Now if we take a look at the controller [C](s), we can see that it's a type II compensator made out of just [on]e op-amp and a few resistors and capacitors as illustrated in [Fig]ure 1.3.

[T]he inverting and non-inverting inputs of the op-amp are effec[tivel]y the summing junction as seen in the feedback control system [block]s. Again, don't get too caught up with understanding what's [happe]ning. The takeaway is that the analog control systems rely

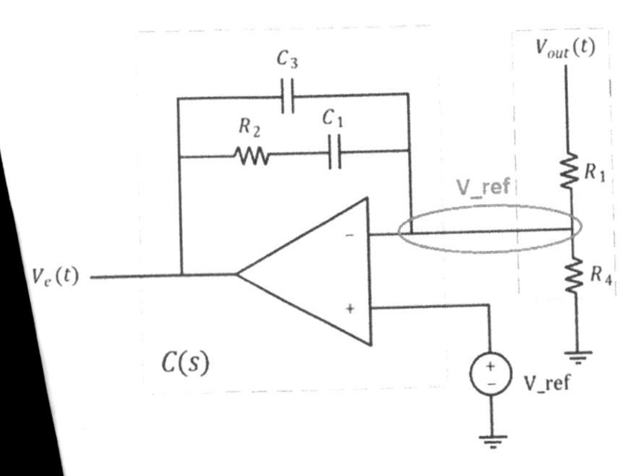

[T]he type II compensator (adds two poles and one zero into

CHAPTER 1

Digital Control with Arduino

1.1 INTRODUCTION

In this chapter, we'll be showing you how you can design and implement a digital controller on the Arduino to control or regulate a continuous-time system. First, we'll start off by discussing the advantages of using digital control as opposed to the traditional analog control and will then briefly go over the Z-transform and discretization of continuous-time systems. Then we'll move on to the six steps for designing and implementing a digital controller on the Arduino and lastly, we'll finish with designing a digital controller for a direct current (DC) motor speed control system. So, this chapter outline is as follows:

- Advantages of digital control
- Z-Transform
- Discretization of continuous-time systems (Tustin's method)
- Implementation of discrete systems on the Arduino

DOI: 10.1201/9781003541356-1

- Digital controller design (Six steps)
- DC motor speed control example: obtaining the transfer function
- DC motor speed control example: controller design
- DC motor speed control example: Arduino implementation

Now keep in mind like we said in this chapter's description this is not an introductory course on signal processing or controls. If this is your first-time hearing words such as Laplace and Z-transforms or sampling stability, then you should first seek out this information elsewhere. This chapter is a practical tutorial series made to bridge the gap between the theory and practice of discrete-time systems and control theory. Again, our goal is to give you the tools needed to design a controller, how you will integrate this into your design is 100% your responsibility to figure out. Therefore, the disclaimer for this chapter is as follows.

- This is NOT an introductory course on LINEAR SYSTEMS, CONTROL THEORY, DIGITAL SIGNAL PROCESSING, ELECTRONICS, ETC.
- This is a practical tutorial course that will allow students, engineers, and hobbyists to apply their known theoretical knowledge of control theory and discrete-time systems.
- We will not hold ourselves responsible for any explanation of the basic fundamentals of linear systems and signals, Arduino coding, and basic electronics. It is required you have some exposure to these topics coming into reading this chapter. So, let's get on with this chapter.

1.2 ADVANTAGES OF DIGITAL CONTROL

Before we discuss why digital control is useful, let's take a look at analog control or continuous-time controllers. Here we have a

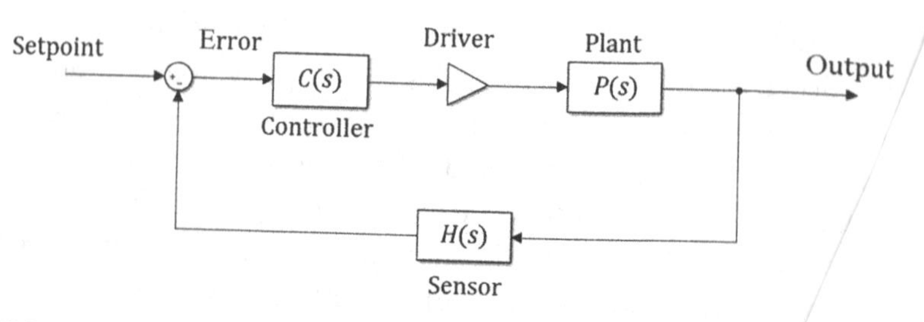

FIGURE 1.1 Analog or continuous-time control system.

system block diagram of a control system where $C(s)$ is controller driving our plant, $P(s)$, to a desired set poi in Figure 1.1.

Now decades ago, even sometimes today $C(s)$ using purely discrete analog components typicall operational amplifiers. Therefore, we can say the back systems (decades ago) relied on analog co pensators for closed-loop control. Wherea systems, they were mainly done using op-am of this is a switch mode power supply topolo converter as depicted in Figure 1.2.

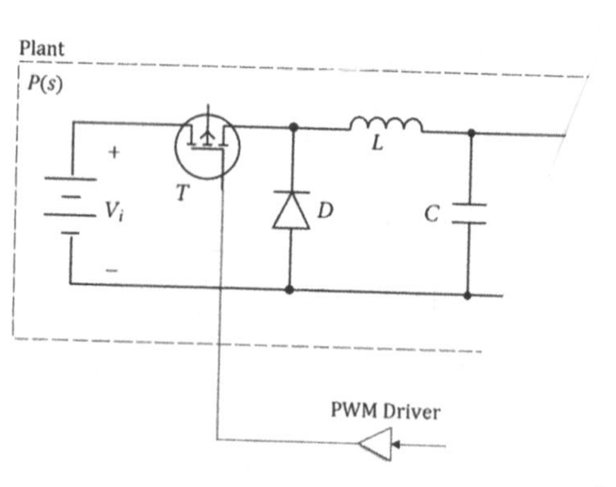

FIGURE 1.2 The buck conver example.

only on analog components for their controller implementation made entirely out of analog components, forming a 100% continuous-time coupled control loop and thus there are no time delays, sampling delays, and quantization errors and they are also easier to analyze and design using classical control techniques such as root locus or Bode plot analysis. Now that's all great. However, again why use digital control if analog controls seem so sufficient? Well with the use of analog control come some disadvantages. First, they're mainly composed of discrete analog components so if you wanted to tweak your controller or play around with various designs, this would mean physically replacing the components which is inconvenient if the components are soldered onto a board or not immediately available. Second, the passive components such as resistors and capacitors degrade and vary in value over time and temperature and thus the transfer function of your controller is never fixed, and the worst-case scenario is that if they are somehow damaged due to environmental temperatures, their values could have drifted so far off that they cause the controller to go unstable. Some systems incorporate multiple control loops so the more controllers there are the more space is required on a circuit board, and this can lead to bulkier designs and also increase the costs of the implementation. Some systems are also time variant or non-linear such that they require an adaptive or a non-linear controller. So, the controller's parameters will change in accordance to the variation of the system in order for the control loop to stay stable or robust. However, this is not possible with analog controllers because the parameters set by discrete components are hard wired and fixed. Therefore, the disadvantages of analog/continuous-time controllers are summarized as the following:

- Since they're composed of analog components (resistors, capacitors, op-amps, etc.), the parameters are fixed.
- Analog components degrade over time and temperature, thus causing the overall transfer function to vary unintentionally.

- Some feedback systems require multiple control loops at high orders; this leads to requiring more board space and bulky designs (multiple op-amps and passives), and overall higher cost.
- For systems whose dynamics are time-variants or non-linear, an adaptive or a non-linear controller may be required, and this is very difficult to achieve using discrete analog components considering the components are hard-wired (soldered) and fixed before deployment.

So, if we go from an analog control scheme to a digital control scheme, you'll notice that we introduce a couple of new blocks. Unlike in the analog control scheme, the feedback signal is now sampled discreetly and fed into a microcontroller and then after the error signals are generated, a discrete control algorithm converts its error signal to a control signal which then leaves a microcontroller to drive the plant as shown in Figure 1.4.

The $C(z)$ is our discrete equivalent of our analog controller $C(s)$ and can be implemented through firmware or embedded code and therefore we can easily adjust the coefficients that define the discrete transfer function $C(z)$ and thus have more flexible control over our feedback system's behavior. So here are the main advantages of digital control. The first advantage is that it allows a designer or a control engineer to readily adjust the

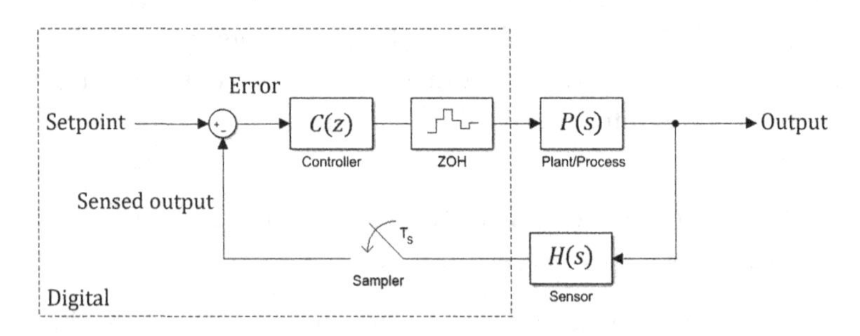

FIGURE 1.4 The digital control scheme blocks.

control parameters through firmware which is inexpensive, saves time, and provides a flexible degree of tuning. Second, since digital controllers are typically implemented on a controller or microprocessor, they take up less board space and allow you to implement multiple control loops and add various features such as saturation or hysteresis with just a few lines of code, and, last, they allow for adaptive or non-linear control laws for systems that are time varying or non-linear in their transient or steady-state behavior. Therefore, the advantages of digital control can be summarized as follows:

- Allows the designer or control engineer to readily adjust the controller's parameters via firmware – inexpensive, saves time, and provides a flexible degree on tuning.
- Since digital controllers are typically implemented on a microcontroller or microprocessor, they take up less board space, allowing you to implement multiple control loops and add various features such as saturation and hysteresis with just a few lines of code.
- Allows for adaptive or non-linear control laws for systems that are time variant or non-linear in their transient or steady-state behavior.

So, this was just a brief motivation for why you should learn to implement digital controllers and their advantages over analog controllers. The rest of this chapter will focus on bridging the gap between the theory and the practical implementation in much greater detail.

1.3 OVERVIEW OF THE Z-TRANSFORM

In this section, we will be briefly looking over the Z-transform and its role in the digital controller's implementation. In continuous-time feedback systems, we represent our controller's transfer function $C(s)$ in the s-domain which is obtained by taking the Laplace

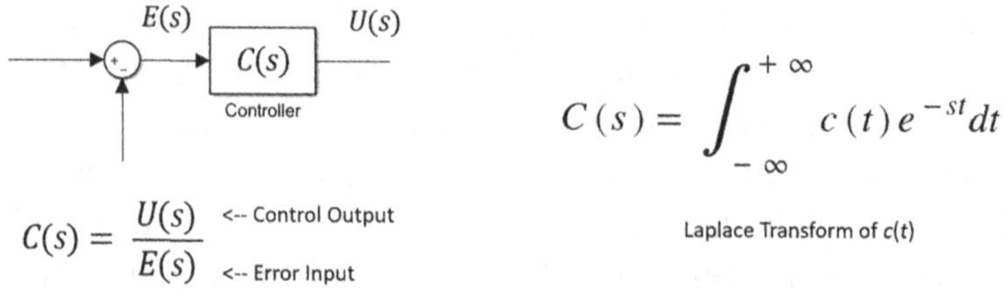

E(s) U(s)

C(s)

Controller

$$C(s) = \frac{U(s)}{E(s)}$$ <-- Control Output / <-- Error Input

$$C(s) = \int_{-\infty}^{+\infty} c(t)e^{-st}dt$$

Laplace Transform of *c*(*t*)

FIGURE 1.5 The Laplace transform of the controller.

transform of the impulse response $c(t)$, where s is a complex number, $\sigma + j\omega$, and $\omega = 2\pi f$. The famous Laplace transform, $L\{\}$, takes a signal or system in the time domain and brings it to its complex frequency-domain representation. Doing so, it converts differential equations to algebraic expressions and the convolution operation does multiplication and helps make analyzing systems easier as equations below and as shown in Figure 1.5.

$$C(s) = \int_{-\infty}^{+\infty} c(t)e^{-st}dt \tag{1.1}$$

$$C(s) = \frac{U(s)}{E(s)} \tag{1.2}$$

where $C(s)$ is the Laplace transform of $c(t)$, $U(s)$ is the control output, and $E(s)$ is the error input.

In other words, it makes analyzing linear systems and designing controllers a lot easier. In the continuous-time or analog-domain, the controller $C(s)$ is derived from a differential equation that describes the dynamics of the physical system where in most cases, it is the dynamics of an electrical analog circuit. For example, the type II compensator used for power supply regulation has the following transfer function as shown in Figure 1.3:

$$C(s) = \frac{V_e(s)}{V_{\text{out}}(s)} = -\frac{1 + C_1R_2s}{(R_2C_1 + R_1C_3)s + R_1C_1C_3R_2s^2} \tag{1.3}$$

To get the transfer function, we can apply the Kirchhoff's Current Law (KCL) on the red node in Figure 1.3. The inverting and non-inverting inputs of the ideal op-amp should have equal voltage, and the current flowing to the inverting point of ideal op-amp is equal to zero; therefore, the red node voltage is also equal to V_{ref} as depicted in Figure 1.3:

$$\frac{V_{ref}-V_e}{\frac{1}{C_3 s}}+\frac{V_{ref}-V_e}{\frac{1}{C_1 s}+R_2}+0+\frac{V_{ref}}{R_4}+\frac{V_{ref}-V_{out}}{R_1}=0 \tag{1.4}$$

$$\frac{V_e}{V_{out}}\Big|_{V_{ref}=0}\rightarrow V_e\left(C_3 s+\frac{1}{\frac{1}{C_1 s}+R_2}\right)=-\frac{V_{out}}{R_1} \tag{1.5}$$

$$\frac{V_e}{V_{out}}\Big|_{V_{ref}=0}=-\frac{1+C_1R_2 s}{\left(R_2C_1+R_1C_3\right)s+R_1C_1C_3R_2 s^2} \tag{1.6}$$

Now don't worry too much about how this compensation was derived. Just understand the concept as understanding the specific controller is way beyond the scope of this chapter. So, we cross multiply the terms, so that they are written as such:

$$-V_{out}(s)\left(1+C_1R_2 s\right)=V_e(s)\left(\left(R_2C_1+R_1C_3\right)s+R_1C_1C_3R_2 s^2\right) \tag{1.7}$$

Then, after taking the inverse Laplace transform or in other words going back to the time domain, we get the following differential equation:

$$-V_{out}(t)-C_1R_2\frac{dV_{out}(t)}{dt}=\left(R_2C_1+R_1C_3\right)\frac{dV_e(t)}{dt}+R_1C_1C_3R_2\frac{d^2V_e(t)}{dt^2} \tag{1.8}$$

Now keep in mind that when we take the Laplace transform or the inverse Laplace transform, you don't actually have to solve in the integrals. You just look at the Laplace transform table provided. So, this differential equation (1.8) is what describes physically the relationship between the input and the output of the compensator. Now this dynamic is implemented using discrete components such as op-amp resistors and capacitors. So, how do we implement this type of dynamic digitally on a microcontroller for example? In order to answer that, we must briefly go over the Z-transform. Just like how the Laplace transform puts a continuous signal in terms of its 's' or its complex frequency domain. The Z-transform puts a discrete or digitized signal in terms of a 'z' or complex frequency domain or '*z*' is equal to '*e*' the power of 'sT' (e^{sT}). Now, $C(z)$ is the Z-transform of $c[k]$, where $c[k]$ is a discrete-time signal sampled at our sampling time, T, as illustrated in Figure 1.6.

So, the Z-transform is essentially the discrete version of the Laplace transform and it follows similar properties of the Laplace transform and it is a mathematical operation that takes a discrete-time signal and converts it to its complex frequency-domain representation:

$$C(z) = \sum_{k=0}^{\infty} c[k] z^{-k} \tag{1.9}$$

Where $z = e^{sT}$

Most importantly it converts convolution in the discrete-time domain to multiplication in the complex frequency domain.

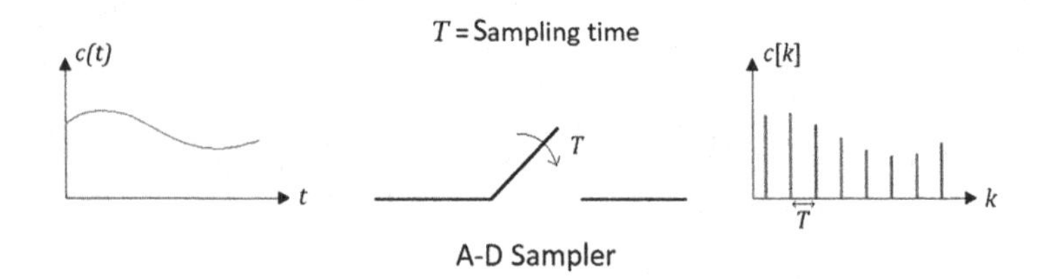

FIGURE 1.6 The Z-transform.

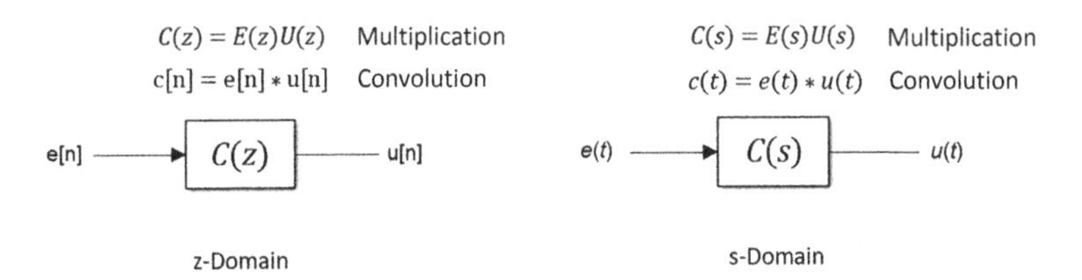

FIGURE 1.7 The Z-transform as a Laplace transform for discrete signals sampled at T.

The Z-transform is essentially the 'Laplace transform' for discrete signals sampled at a fixed sample rate T as given in the following equations and is shown in Figure 1.7:

$$c[n] = e[n] * u[n] \rightarrow C(z) = E(z)U(z) \tag{1.10}$$

$$c(t) = e(t) * u(t) \rightarrow C(s) = E(s)U(s) \tag{1.11}$$

It is also essential to know the mapping between the s-plain and the z-plane which comes from the following expression:

$$z = e^{sT} \tag{1.12}$$

$$s = \sigma + j\omega \tag{1.13}$$

Now σ is a real component and ω is the imaginary component. It is apparent from this formula that the left half plane of the s-plane maps to inside the unit circle as shown in Figure 1.8, where a point can be represented by its magnitude, $e^{\sigma T}$, and its angle, ωT, as represented in the following equation:

$$z = e^{sT} = e^{\sigma T} e^{j\omega T} = e^{\sigma T} \angle \omega T \tag{1.14}$$

So, if there was a point on the origin of the s-plane, it would map to 1 on the real axis of the z-plane since the magnitude is 1

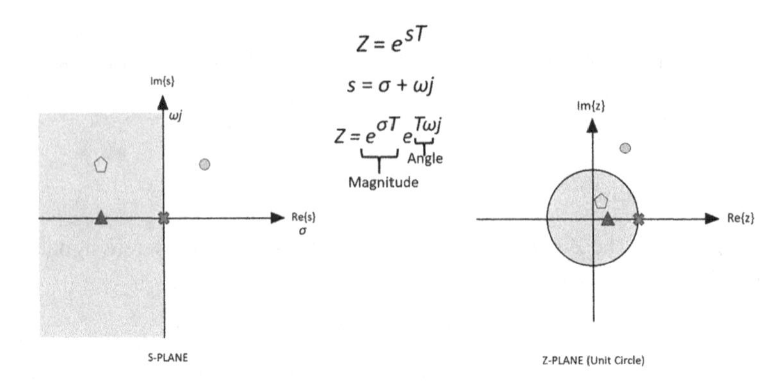

FIGURE 1.8 Mapping between s-plane and z-plane.

and the angle is 0, and if we were to move further negative in the real s-plane, this would correlate to moving closer to the origin of the z-plane However, if we move to the right of the s-plane, this would correlate to a magnitude greater than 1 and, thus, we would be outside the unit circle, and since we also have an imaginary component, the angle is non-zero. Now, if we keep the imaginary component the same and move the real component to the left half-plane, we will find ourselves back inside the unit circle in the z-plane, and, however, the angle will remain the same as represented in equations below and as shown in Figure 1.8:

$$z = e^{sT}\,\big|_{s=0} = 1\angle 0^{o} \tag{1.15}$$

$$z = e^{sT}\,\big|_{s\ll 0} = \left(e^{\sigma T} \ll 1\right) \rightarrow 0\angle \omega T \tag{1.16}$$

$$z = e^{sT}\,\big|_{s>0} = (e^{\sigma T} > 1)\angle \omega T \tag{1.17}$$

This is too crucial to the digital controller design process but it's good to know how the routes match between the Laplace and the Z-transform. A stable system should have all its poles within the unit circle. A pole outside the unit circle would indicate an

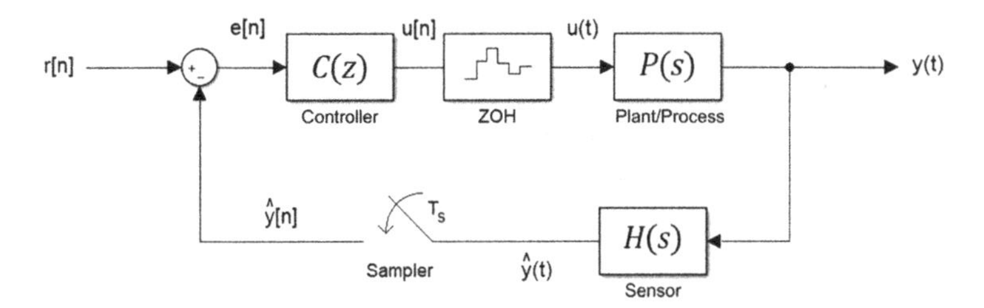

FIGURE 1.9 Closed-loop feedback diagram of digital control.

unstable system. So, taking a look at our closed-loop feedback diagram, a controller $C(z)$ is now digital and is represented in the z-domain preceding it as a sampler which is sampling at a sampling time of T_S and after it comes our zero-order hold (ZOH) block. We're just theoretically bringing our discrete signal back to the continuous-time domain to drive the plant as illustrated in Figure 1.9.

Now, we'll go more into both of these blocks in later sections of this chapter. So, let's say we've designed $C(z)$, that is great. Now how do we implement this? To answer that let's say $C(z)$ equals the following transfer function which you can clearly see relates are discrete error input $E(z)$ and our controller's discrete output $U(z)$ as below equation:

$$C(z) = \frac{U(z)}{E(z)} = \frac{b_0 + z^{-1}b_1 + z^{-2}b_2 + \ldots + z^{-M}b_M}{a_0 + z^{-1}a_1 + z^{-2}a_2 + \ldots + z^{-M}a_M} \tag{1.18}$$

Now, if we cross multiply the turns, we get the following expression:

$$\begin{aligned} &U(z)(a_0 + z^{-1}a_1 + z^{-2}a_2 + \ldots + z^{-M}a_M) \\ &= E(z)(b_0 + z^{-1}b_1 = +z^{-2}b_2 + \ldots + z^{-M}b_M) \end{aligned} \tag{1.19}$$

Now if we distribute $U(z)$ and $E(z)$ on the respective sides of the equation and then we take the inverse Z-transform of the expression, or in other words going back to the discrete-time domain, we get the following difference equation:

$$Z^{-1}\{X(z)\} = \frac{1}{2\pi j}\oint X(z)z^{n-1}dz \tag{1.20}$$

$$\begin{aligned} a_0u[n]+a_1u[n-1]+a_2u[n-2]+\ldots+a_Mu[n-M] \\ = b_0e[n]+b_1e[n-1]+b_2e[n-2]+\ldots+b_Me[n-M] \end{aligned} \tag{1.21}$$

This is called a difference equation which is essentially the discrete version of the differential equation and this equation is what is coded and used to develop the code in the firmware to implement a digital controller in a real-time system. Now, notice the '*a*' and '*b*' coefficients in front of each discrete-time term in the differential equation case. These had to be using resistors and capacitor values because this is a difference equation and they can be implemented digitally. So, these coefficients can simply be adjusted in the code. Therefore, as the title of this section indicates, this is a very brief overview of the Z-transform; in the later sections, we will dive much deeper into making intuitive sense out of it and you'll see where it's relevant in the digital controller design process.

1.4 DISCRETIZATION OF CONTINUOUS-TIME SYSTEMS

In this section, we'll be going over the discretization of analog or continuous-time systems using Tustin's method. Whenever we have a pure continuous-time feedback loop with the continuous-time plant, sensor, and controller, we design our controller, $C(s)$, using classical control techniques as shown in Figure 1.1. The two most popular techniques being root locus and the frequency response or, Bode plot analysis. Using the root locus, we observe how close loop poles move in the s-plane as a function of the gain of our control or $C(s)$. Using the frequency response analysis technique, we observe the Bode plot of the entire feedback loop and check the gain and phase margins to assess the stability and the transient performance. So, if a continuous-time control or $C(s)$ is designed using these classical control techniques, how do we go about designing a digital controller, $C(z)$, to drive a continuous-time plant such as a power supply, a motor, or a heater system. The best way to derive a digital controller is to first derive a continuous controller $C(s)$ that can appropriately drive our plant and then knowing that z equals e to the sT ($z = e^{sT}$), we can convert controller transfer function, $C(s)$, to a discrete controller or $C(z)$ such that it mimics the continuous-time controller's behavior in the discrete/digital domain in regard to its frequency response as you can see in Figure 1.4. So essentially our goal is to design $C(z)$ such that its frequency response matches closely with $C(s)$ frequency response. So, let's say we have a system or plant that we want to control with feedback and we design the following continuous-time controller or controller $C(s)$ that has a pole in −5 and zero at −1 as represented in the following equation:

$$C(s) = \frac{s+1}{s+5} \tag{1.22}$$

So, we know that z equals e to the sT as below:

$$z = e^{sT} \tag{1.23}$$

If we arrange in the equation the 's' as a function of 'z', we get that s equals a natural logarithm of z over T, therefore, we have the following equation:

$$s = \frac{\ln(z)}{T} \tag{1.24}$$

So, substituting for 's' in $C(s)$, we get $C(z)$ that equals natural logarithm of 'z' over T plus 1 over natural logarithm of 'z' over T plus 5 as follows:

$$C(z)=\frac{\frac{\ln(z)}{T}+1}{\frac{\ln(z)}{T}+5} \tag{1.25}$$

This equation is suddenly a non-linear equation due to the natural logarithm term and therefore we cannot take the inverse Z-transform of this, hence, we cannot work with this equation. So, what we'll need to do instead is approximate z equals e to the sT (e^{sT}) as a first-order linear equation. There are many approximation methods that have been proposed but one in particular does a better job than the rest, and that's Tustin's method which is also known as a bilinear transformation as given in the following equations:

$$\text{Euler's method}: z=e^{sT}\approx 1+sT \tag{1.26}$$

$$\text{Backward difference}: z=e^{sT}\approx \frac{1}{1-sT} \tag{1.27}$$

$$\text{Tustin's method (bilinear transformation)}: z=e^{sT}\approx \frac{1+s\frac{T}{2}}{1-s\frac{T}{2}} \tag{1.28}$$

Now, if we map the left half plane of the s-plane to the z-plane using each of these approximations, you can see that Euler's method and the backward difference methods map very poorly in comparison to Tustin's method as illustrated in Figure 1.10.

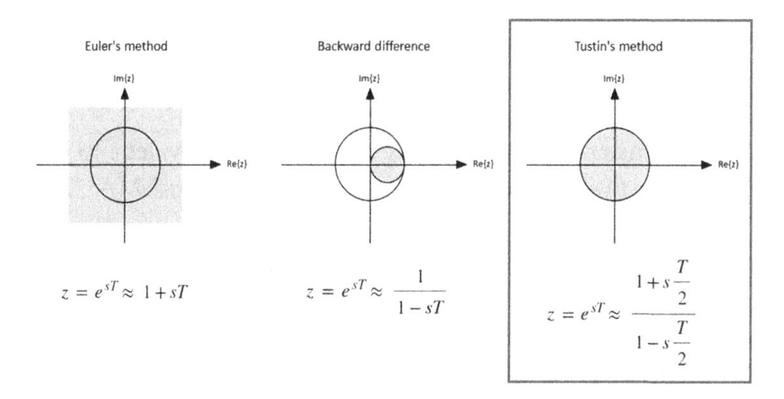

FIGURE 1.10 Mapping of the left-hand s-plane (Re{s}< 0) to the z-plane (unit circle).

Therefore, using Tustin's method, the mapped z-plane for values Re{s} < 0 matches the closest to the original z-plane of $z = e^{sT}$. So, if we want our digital controller's frequency response to match very closely to the frequency response of our continuous-time controller, it is in our best interest to go with Tustin's method. So, if we rearrange this new approximation of 's' as a function of 'z', we get that 's' equals 2 over T times Z minus 1 over Z plus 1 as the following equation:

$$z = e^{sT} \approx \frac{1 + s\frac{T}{2}}{1 - s\frac{T}{2}} \rightarrow s = \frac{2}{T}\frac{z-1}{z+1} \tag{1.29}$$

If we now substitute for 's' and $C(s)$ and, for example, set the sampling time T to be 0.02 seconds, we get the following z-domain transfer function:

$$C(s) = \frac{s+1}{s+5}\Bigg|_{\substack{s = \frac{2}{T}\frac{z-1}{z+1} \\ T(\text{sampling time}) = 0.02\,\text{seconds}}} \tag{1.30}$$

$$C(z) = \frac{0.9619z - 0.9429}{z - 0.9048} \tag{1.31}$$

This transfer function $C(z)$ is the discrete equivalent to the continuous-time transfer function $C(s)$. Now we can use MATLAB to verify that their frequency responses match close enough. So, let's verify with MATLAB that $C(z)$ has similar frequency response as $C(s)$. We'll use the code below, where the first code line is declaring our continuous-time transfer function $C(s)$ and then to get the discrete transfer function $C(z)$, reuse the 'c2d' function with the method set to 'tustin' and the sampling time, Ts, set to 0.02.

```
C_s = tf([1 1],[1 5]);     %% C(s) is the transfer
                              function
Ts = 0.02;                 %% Sampling time is 0.02 s
C_z = c2d(C_s,Ts,'tustin');  %% Discretizing C(s)
                                using Tustin's
                                method
bode(C_s);                  %% Plotting the frequency
                               response C(s)
hold on;
bode(C_z);                  %% Plotting the frequency
                               response C(z)
```

Then, we're plotting the Bode plots for both $C(s)$ and $C(z)$ on the same plot to see how they overlap. Now after running the code on MATLAB, we will get the figure showing the frequency responses of both $C(s)$ and $C(z)$, and they are matched very closely with one another as shown in Figure 1.11.

Although they do not perfectly align, in our case of designing a digital controller that behaves like the continuous one, it'll do just fine. Hopefully, this section gave you enough information to know how to discretize a continuous-time system. In the next section, we will look into how a digital system like the one derived here is implemented in code on the Arduino Integrated Development Environment (IDE).

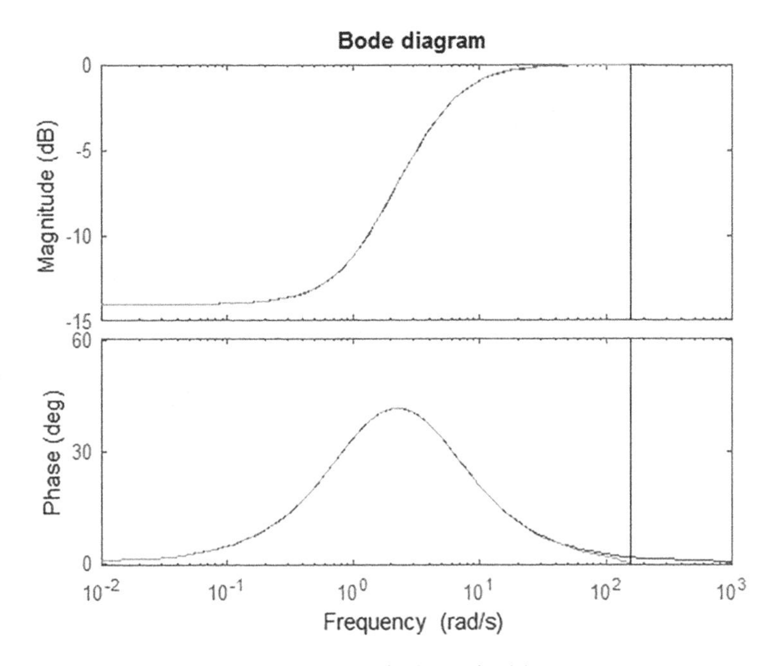

FIGURE 1.11 The Bode diagram of $C(s)$ and $C(z)$.

1.5 ARDUINO IMPLEMENTATION OF DISCRETE-TIME SYSTEMS

In this section, we will go over to see how you can implement a discrete-time system on the Arduino. Now we're going to step back from any discussion about feedback control and just see how we can implement any arbitrary discrete-time system on a micro-controller so let's say we have a second-order system, $H(s)$, that we wish to discretize as below:

$$H(s) = \frac{20}{s^2 + 2.5s + 20} \tag{1.32}$$

Now, if we apply a step input for this system which we can simulate with MATLAB using the following code we can observe how the output behaves and we get a very typical under-damped second-order response as shown in Figure 1.12:

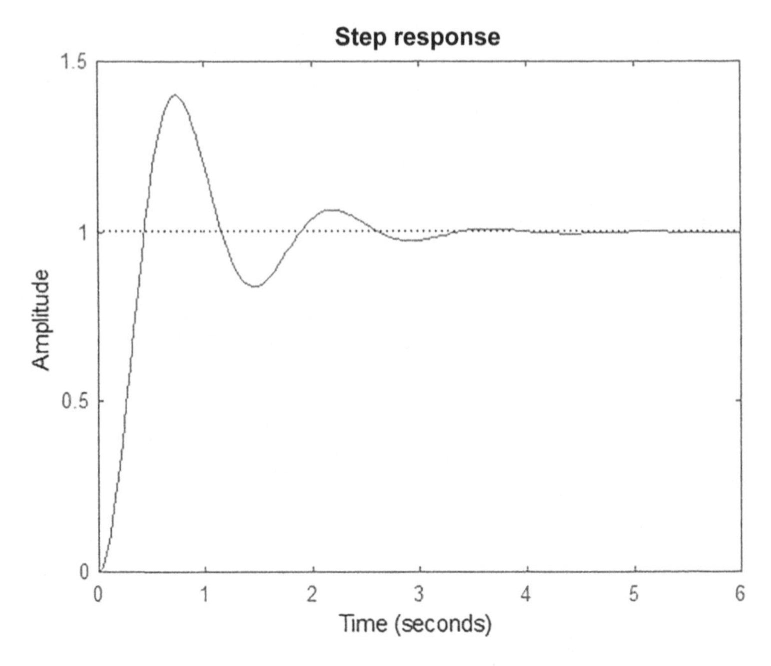

FIGURE 1.12 The step response of second-order continuous-time system, $H(s)$.

```
H_s = tf([20],[1 2.5 20]);
step(H_s);
```

We can also observe the frequency response of the system using the 'bode' function on MATLAB by the following code:

```
H_s = tf([20],[1 2.5 20]);
bode(H_s);
```

In the Bode plot as depicted in Figure 1.13, we will have a natural or resonance frequency of about 4.07 rad/s.

So, it's a pretty slow and low bandwidth system and because we want to discretize this to implement on Arduino, we use the 'c2d' function on MATLAB as shown in the previous section and we

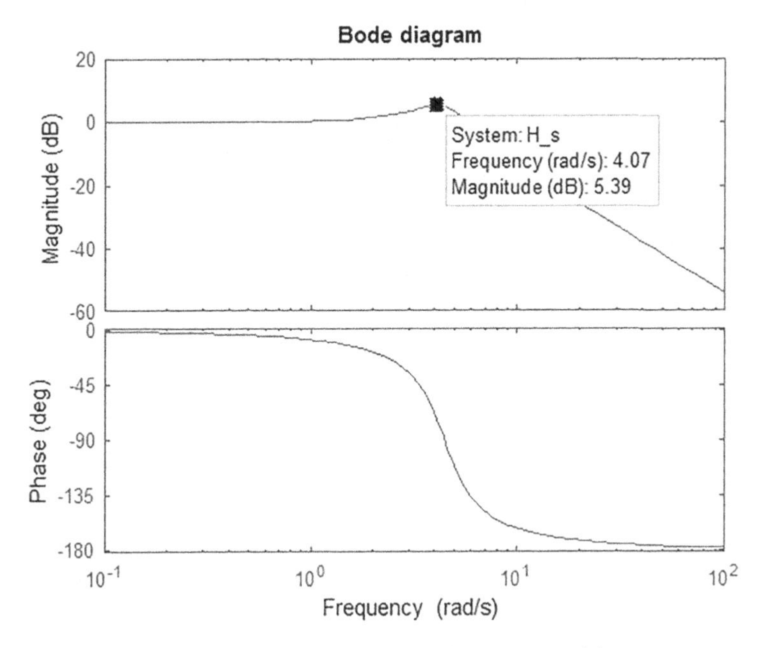

FIGURE 1.13 The frequency response of the system, $H(s)$.

set the method to 'tustin' and the sampling-time to T which is equal to 0.1 seconds as the following code:

```
H_s = tf([20],[1 2.5 20]);
T = 0.1;
H_z = c2d(H_s,T,'tustin');
```

This is the discrete transfer function we get as the equation below:

$$H(z) = H(s)\Big|_{\substack{s=\frac{2}{T}\frac{z-1}{z+1} \\ T=0.1\,\text{sec}}} = \frac{0.04237z^2 + 0.08475z + 0.04237}{z^2 - 1.61z + 0.7797} \tag{1.33}$$

We'll call this discrete-system, $H(z)$, and if we plot the transient response of both the analog and digital systems, you can see that

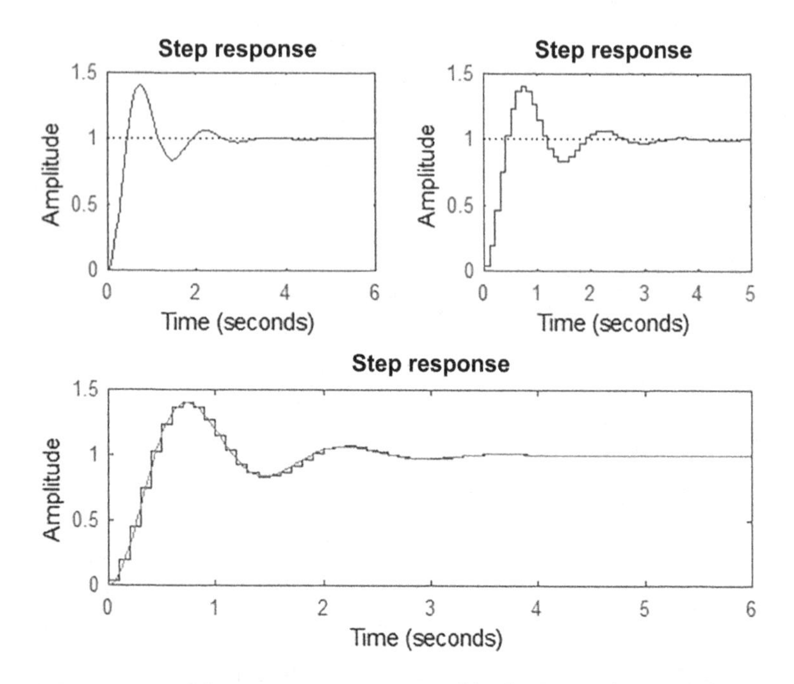

FIGURE 1.14 The transient response of both the analog and digital systems.

their responses line up with each other as shown in Figure 1.14, and the MATLAB code is as below:

```
H_s = tf([20],[1 2.5 20]);
T = 0.1;
H_z = c2d(H_s,T,'tustin');
subplot(2,2,1);
step(H_s);
subplot(2,2,2);
step(H_z);
subplot(2,2,[3,4]);
step(H_z, H_s);
```

Also, if we plot the frequency responses of both systems, we can see that they line up relatively well with some error at the higher frequency as shown in Figure 1.15, and it should be said that when

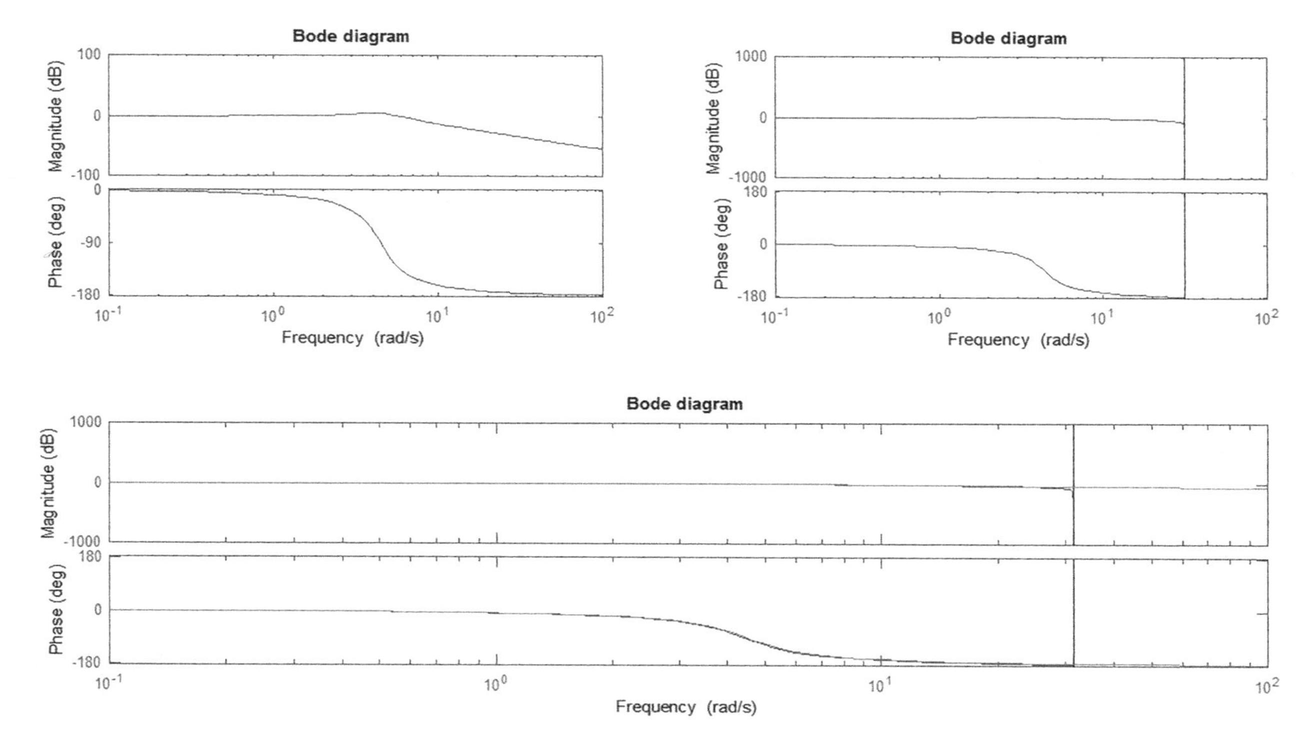

FIGURE 1.15 The frequency response of both the analog and digital systems.

it comes to design the controllers and digitizing systems, it isn't necessary that you get a controller that behaves exactly 100% like its analog equivalent.

The following MATLAB code is used for plotting the frequency responses of both systems:

```
H_s = tf([20],[1 2.5 20]);
T = 0.1;
H_z = c2d(H_s,T,'tustin');
subplot(2,2,1);
bode(H_s);
subplot(2,2,2);
bode(H_z);
subplot(2,2,[3,4]);
bode(H_z, H_s);
```

The goal is to design a controller that will robustly control the open-loop system. Now, we want to write this transfer function in its causal form with the following equation:

$$H(z)=\frac{0.04237z^2+0.08475z+0.04237}{z^2-1.61z+0.7797} \tag{1.34}$$

So, because it's a second-order system, we divide the numerator and the denominator by z^2 to get the following causal expression:

$$H(z)=\frac{Y(z)}{X(z)}=\frac{0.04237+0.08475z^{-1}+0.04237z^{-2}}{1-1.61z^{-1}+0.7797z^{-2}} \tag{1.35}$$

Then as before we cross multiply the terms to separate $X(z)$ and $Y(z)$ as below:

$$\begin{aligned}&Y(z)\left(1-1.61z^{-1}+0.7797z^{-2}\right)\\&\quad=X(z)\left(0.04237+0.08475z^{-1}+0.04237z^{-2}\right)\end{aligned} \tag{1.36}$$

$$Y(z)-Y(z)1.61z^{-1}+Y(z)0.7797z^{-2}$$
$$=X(z)0.04237+X(z)0.08475z^{-1}+X(z)0.04237z^{-2} \quad (1.37)$$

Also, after distributing $X(z)$ and $Y(z)$ on both sides, respectively, and then taking the inverse Z-transform to go back to the discrete-time domain, we get the following expression:

$$y[n]-1.61y[n-1]+0.7797y[n-2]$$
$$=0.04237x[n]+0.08475x[n-1]+0.04237x[n-2] \quad (1.38)$$

What you'll notice over and over again is that z^{-1} raised to the −1 correlates to the previous sample from one discrete-time step ago and that z^{-2} raised to the −2 correlates to the sample from two discrete-time steps ago. So, we then take our prior output terms and move them over to the right side as the following equation:

$$y[n]=0.04237x[n]+0.08475x[n-1]+0.04237x[n-2]$$
$$+1.61y[n-1]-0.7797y[n-2] \quad (1.39)$$

Now, we have a difference equation that is causal, meaning that its output only relies on the current and the previous input and output values. So, now let's go and implement this difference equation on the Arduino and see it in action. We'll first start off or we do know the script by first declaring variables and we're setting our sampling time to T equals 100 which is one hundred milliseconds or 0.1 seconds in this case. We will then create two variables called last_time and current_time, which we'll use to keep a constant sample time interval, 0.1 seconds, in every iteration, and then we'll have to declare our input and output variables for storing the current and previous values as given by the following code:

```
//Declaring variables
int T = 100;          //sampling time in milliseconds
                      (T = 0.1s)
```

```
unsigned long last_time;
unsigned long current_time;
float x;              //x[n]
float y;              //y[n]
float x_1;               //x[n-1]
float y_1;               //y[n-1]
float x_2;               //x[n-2]
float y_2;               //y[n-2]
```

Also, for our void setup function which runs only once at the beginning of the program, we set all our variables to zero since we want our system response to have a zero initial condition in order to mimic a step response, we'll have to generate a unit step function which is done here in the code below and the waveform of $x[n]$ is illustrated in Figure 1.16:

```
//Test Input Discrete Step Signal
vaoid x_input(){
 if(current_time == 1000){ //After 1000ms or 1
second, step x[n] from 0 to 1
     x = 1.0;
 }
}
```

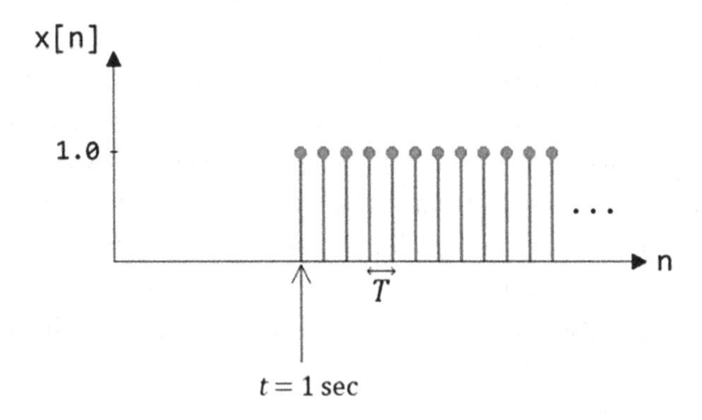

FIGURE 1.16 Testing the input step signal.

The if statement condition is that once the elapsed time has reached 1000 milliseconds or 1 second, the input $x[n]$ should be equal to 1. So essentially, we're going from x equals 0 to x equals 1, the moment we reach 1 second. Here, as illustrated in the code below, is the main void loop where we use a millis() function to keep track of the current elapsed time in milliseconds. We then compute the delta time or time difference between the current time and the previous time, and once this delta time value reaches our sample time of 0.1 seconds, we execute the if statement and then we save our current time as the last time.

```
//Main Loop
void loop() {
 current_time = millis();  //stores the program
 running time(ms)
 int delta_time = current_time - last_time;  //
 computes ΔT
 if (delta_time >= T) { //when ΔT = T, read
 input x[n] and compute y[n]
   elapsed_time = elapsed_time + T/1000.0; //
   compute elapsed time in seconds
   x_input(); //obtain input x[n]
   y = 0.0424*x + 0.0848*x_1 + 0.0424*x_2 +
   1.61*y_1 - 0.7797*y_2;
   x_2 = x_1;     //store x[n-1] as x[n-2] for
   next iteration
   y_2 = y_1;     //store y[n-1] as y[n-2] for
   next iteration
   x_1 = x;       //store x[n] as x[n-1] for
   next iteration
   y_1 = y;       //store y[n] as y[n-1] for
   next iteration
   last_time = current_time;  //restarts ΔT or
   delta_time back to 0
 }
}
```

This method allows us to have artificial control over our sample time interval despite the higher sampling frequency of the Arduino's ADC. So, in our 'if' statement, we will include our step input function, x_input, and then our difference equation. Notice that this line of code is exactly the one we derived before in Equation (1.31), and then lastly, we store the current values as the previous or delayed values for the next iteration, and there you have it as you can see in the above code. That is all you need to implement a discrete-time system on the Arduino. Let's run this code on the Arduino IDE and monitor the input and the output. So here, we have the Arduino code we just discussed.

```
float T = 100.0;
unsigned long last_time;
unsigned long current_time;
double elapsed_time;
double x;
double y;
double x_1;
double y_1;
double x_2;
double y_2;
void setup() {
 Serial.begin(9600);
 x = 0.0;
 y = 0.0;
 x_1 = 0.0;
 y_1 = 0.0;
 x_2 = 0.0;
 y_2 = 0.0;
}
void loop() {
 current_time = millis();
 int delta_time = current_time - last_time;
 if (delta_time >= T) {
   elapsed_time = elapsed_time + T/1000.0;
   x_input();
```

```
    y = 0.0424*x + 0.0848*x_1 + 0.0424*x_2 +
    1.61*y_1 - 0.7797*y_2;
    x_2 = x_1;
    y_2 = y_1;
    x_1 = x;
    y_1 = y;
    Serial.print(elapsed_time);
    Serial.print("      ");
    Serial.println(y);
    last_time = current_time;
  }
}
void x_input() {
  if (current_time == 1000) {
    x = 1.0;
  }
}
```

One neat thing about the Arduino IDE is that you can monitor any value on the Serial Plotter, so we'll want to monitor the step input x and the output y and see how they behave; hence, from Tools menu, we select the Serial Plotter option. So, let's run the program as shown in Figure 1.17, we're getting the slow

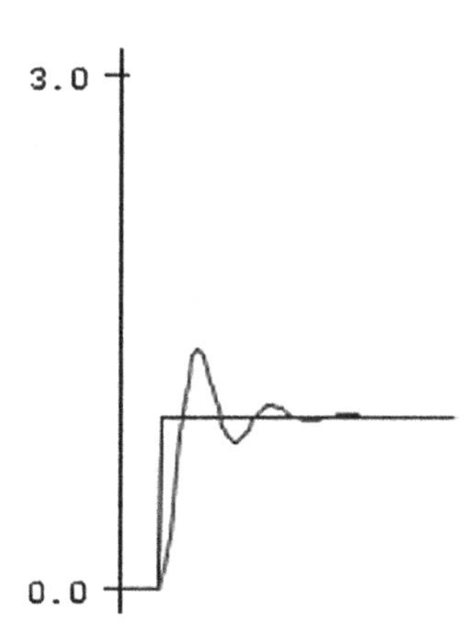

FIGURE 1.17 The slow under-damped response with Arduino IDE's Serial Plotter.

under-damped response that we expected but in order to verify that this result is theoretically correct, let's confirm it with MATLAB Simulink.

We'll first create a discrete transfer function block and type in the numerator and denominator coefficients in Equation (1.34), and then we'll set the sample time to 0.1 seconds as depicted in Figure 1.18.

Next, we'll add the step input block which will step up from 0 to 1 at 1 second with sample time equal to 0.1 seconds as illustrated in Figure 1.19.

Then lastly, we'll add a scope to plot and observe the output and the input and run a 10-second simulation as shown in Figure 1.20.

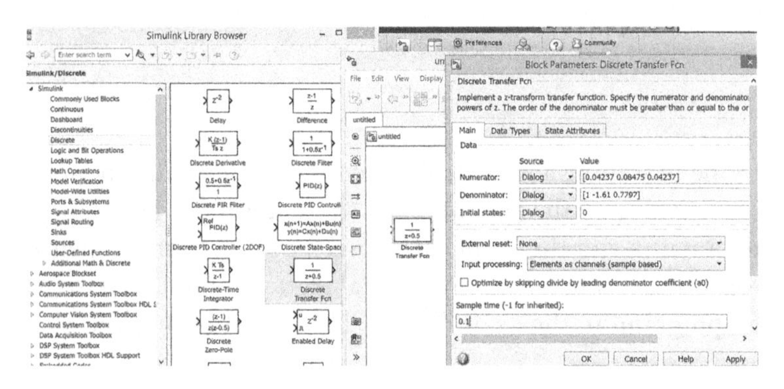

FIGURE 1.18 Setting the discrete transfer function parameters.

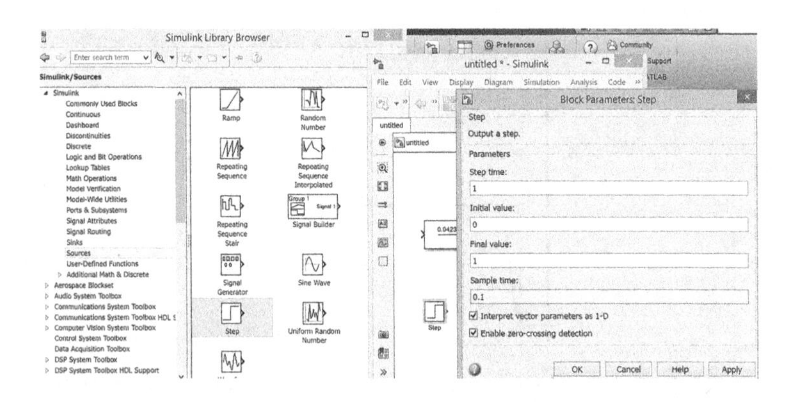

FIGURE 1.19 Adding the step input block.

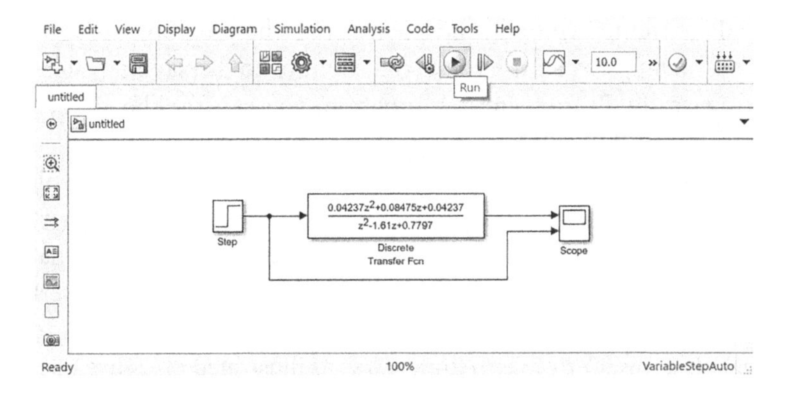

FIGURE 1.20 Adding a scope and running the simulation.

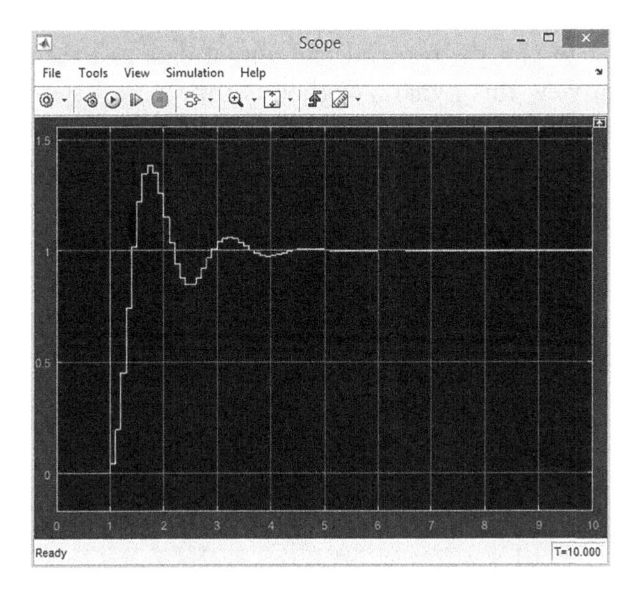

FIGURE 1.21 The discrete transfer function step response in scope.

From the scope, we can see that this is what we expect from the theoretical model of our discrete transfer function as shown in Figure 1.21.

Now we want to compare this with the implementation result. So, we'll need to monitor the time and output data, so that we can create a two-dimensional variable on MATLAB. Now going back to the Arduino, compile and then upload the program from the Tools menu, we can use the Serial Monitor to display the values numerically. By running the program, you can see the elapsed time and output columns on the Serial Monitor.

Currently, we have completed the initial steps. Next, we will copy the values up to 10 seconds into Notepad++ and then import them into MATLAB. To do this, we will create a data variable that includes both the time and output data. Additionally, we will insert a row above the time value of 0.1 and set it to 0, as we want to include a row for time zero (time = 0 s), as illustrated in Figure 1.22.

After that, we do rename the data variable as Output_Data as shown in Figure 1.23.

Then, we use the From Workspace block from Sources in Simulink to plot the experimental output data and then use a scope to view this signal alongside the theoretical step input. Therefore, we do double click on the From Workspace block and set Data as Output_Data and Sample time to 0.1 seconds as depicted in Figure 1.24.

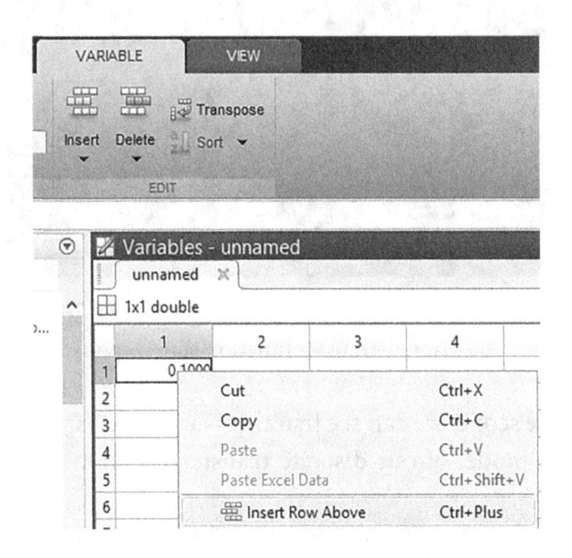

FIGURE 1.22 Creating a data variable that contains time and output data and adding a row above time 0.1 s.

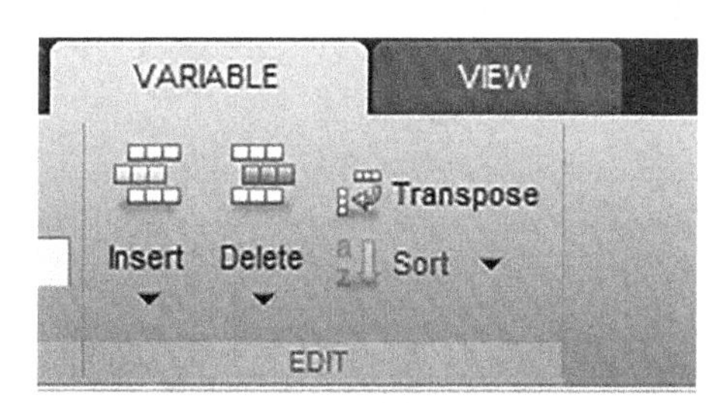

FIGURE 1.23 Renaming the data variable as Output_Data.

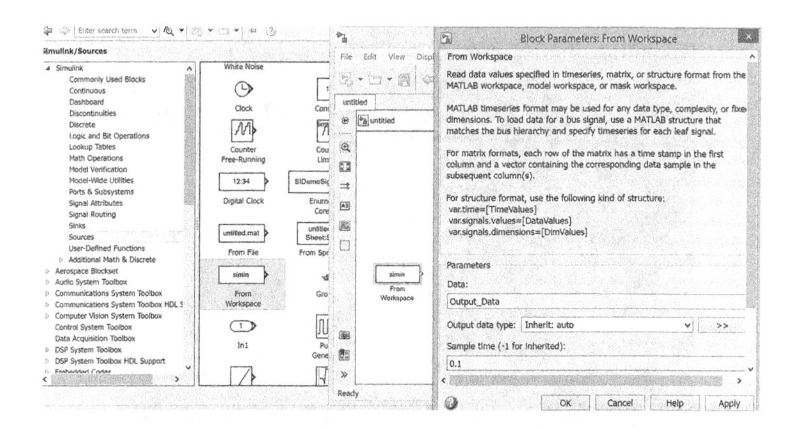

FIGURE 1.24 Setting the From Workspace block properties.

Then we connect it to a scope to compare the theoretical and experimental results, and if we were to plot the theoretical simulated output against the implementation output, we can see that they lie almost exactly on top of one another as shown in Figures 1.25 and 1.26, respectively.

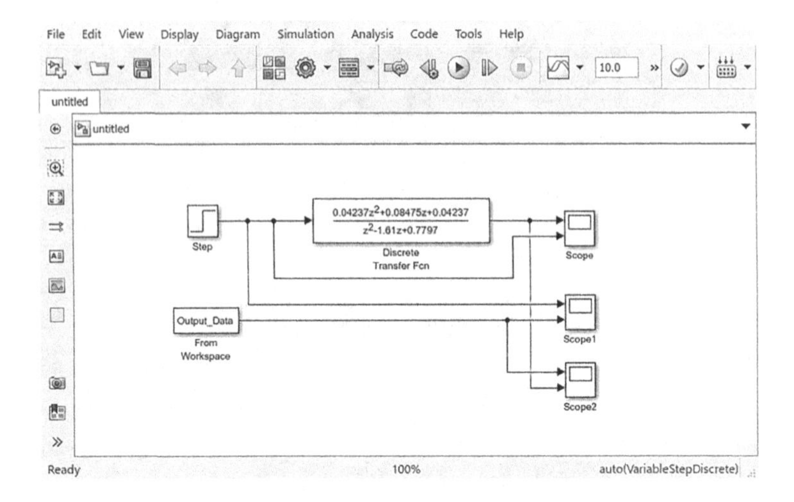

FIGURE 1.25 Theoretical simulated output against the implementation output.

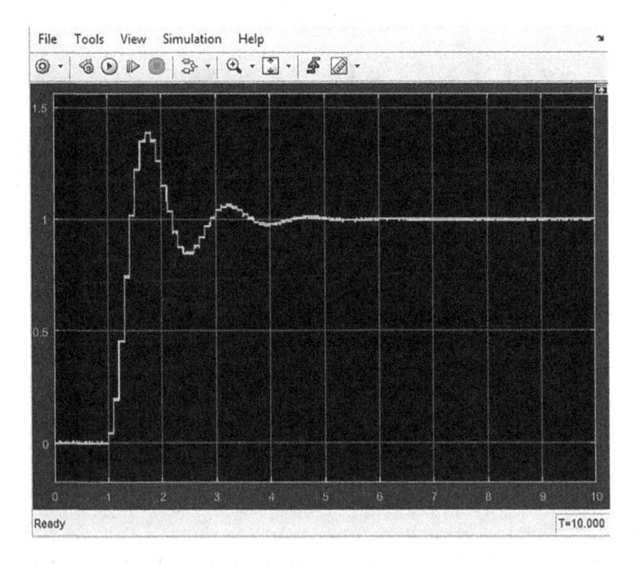

FIGURE 1.26 Exact overlap of theoretical simulated output and implementation output.

This goes to show that you can use the Arduino or any microcontroller to implement a difference equation or discrete-time system. So now that we're confident that we can implement a discrete-time system in the real time. Let's move on to discussing how we can implement the feedback control algorithm on the Arduino.

1.6 DIGITAL FEEDBACK CONTROL ALGORITHM

In this section, we'll build upon what we learned in the previous section and now discuss how to implement feedback control structure on the Arduino. So, in the last section, we learned how to implement a discrete-time system such as $C(z)$ in this case as shown in Figure 1.4. However, how do we code up the feedback control algorithm so that we can use an Arduino in a control loop to digitally regulate a continuous-time plant? Or, how do we implement the digital control scheme inside the dashed line as shown in Figure 1.4? In Figure 1.27, you'll see the portion of the feedback loop we wish to implement digitally and, in the code below, you'll see the main loop block that we discussed in the

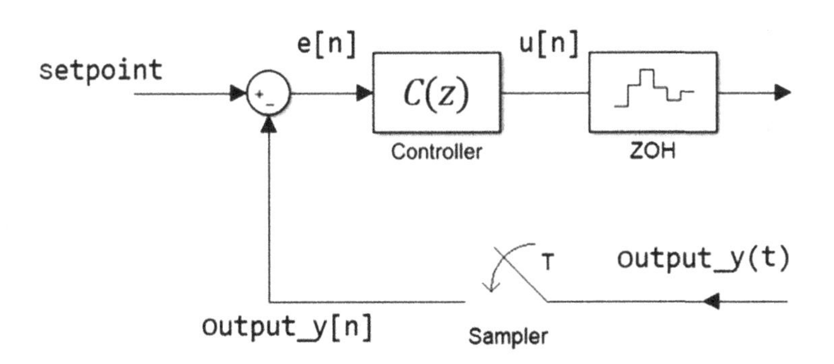

FIGURE 1.27 Digital implementation for the portion of the feedback loop.

previous section for implementing the controller transfer function, $C(z)$, as the following equation:

$$C(z)=\frac{U(z)}{E(z)}=\frac{b_0+b_1z^{-1}+b_2z^{-2}+\ldots+b_mz^{-m}}{1+a_1z^{-1}+a_2z^{-2}+\ldots+a_nz^{-n}} \tag{1.40}$$

```
void loop() {
  current_time = millis();  //stores the program
  running time(ms)
  int delta_time = current_time - last_time;  //
  computes ΔT
  if (delta_time >= T) { //when ΔT = T, read
  input x[n] and compute y[n]
    sensed_output(); //obtain a current sample,
    output_y[n]
    e = setpoint - output_y;  //e[n] = setpoint
    - output_y[n]
    u = -a_1*u_1- … -a_n*u_n + b_0*e + b_m*e_m;
    //C(z)
    u_n = u_(n-1);         //store u[n-(n-1)] as
    u[n-n] for next iteration
    e_m = e_(m-1);         //store e[n-(m-1)] as
    e[n-m] for next iteration
    . . .
    u_1 = u;               //store u[n] as u[n-1]
    for next iteration
    e_1 = e;               //store e[n] as e[n-1]
    for next iteration
    last_time = current_time;  //restarts ΔT or
    delta_time back to 0
  }
}
```

So, let's work our way from the feedback path, so, if you take a look at the 'if' statement in the code, you'll notice that it in itself is the sampler as it only executes every sample time of *T*. So, for example, if *T* equals one second, then the code inside the 'if' statement will iterate or update every one second. That being said if we call a function which may be declared somewhere else in the script and reads in some discrete or analog sensed output data and if we place that in the 'if' statement, then we're artificially sampling the output signal every T second. So, this code so far takes care of the sampling block. Next, we take our sensed output and then subtract it from a set point or a desired output to generate an error signal and this is done by doing a simple subtraction. Although we don't show it here, the variables we're introducing are declared at the beginning of the script. So, after we've obtained our error, we want to generate a control single to drive our plant. So next we implement the controller as a difference equation where the error e is our input and u is our control output. Here in the above code, we have it written as a pseudo code for the fallen transfer function with 'a' and 'b' coefficients in Equation (1.30). If you notice this is exactly the same implementation we covered in the previous section. Nothing is different about the way it is implemented; all that's different is what we're feeding in as our input and getting back as our output. The difference equation structure is exactly the same. So, that is all, and this is all you need to implement the feedback control structure as shown in Figure 1.4. Now, you may wonder what is a zero-order hold (ZOH) block for as you can see in Figure 1.4, and why did we include it? Well let's step back and assess its theoretical function if we have a continuous signal, $x(t)$, and pass it through a sampler which samples every T second, then we theoretically end up with the discrete signal, $x[n]$. Now, what the zero-order hold (ZOH) block does is hold the sample value for T seconds so that we get a staircase looking version of the continuous signal as depicted in Figure 1.28.

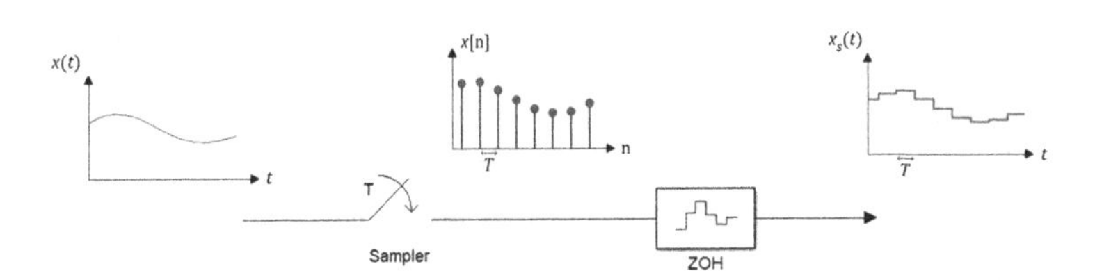

FIGURE 1.28 The zero-order hold (ZOH) block operation.

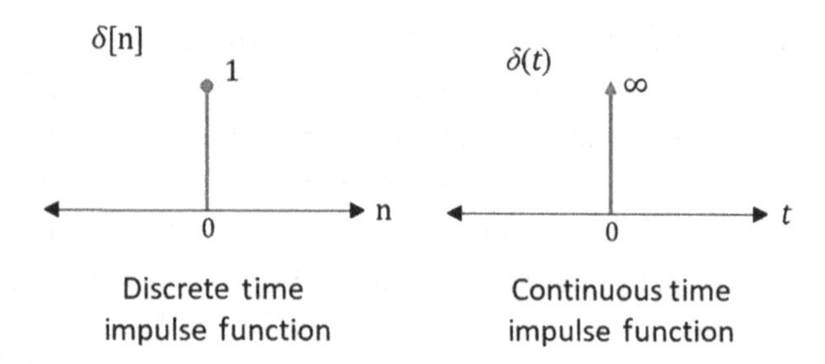

FIGURE 1.29 The impulse function is used only for theoretical derivation and analysis.

So, how do we implement a ZOH block? The answer is we don't have to. It's already intrinsically part of the firmware or code. Keep in mind that this signal $x[n]$ does not exist in our physical reality and it is simply a mathematical representation of a sampled signal. Nowhere, inside the microcontroller, you will find a signal where the value only shows up in for testimony at the beginning of the sample period and then suddenly disappearing in between. This kind of representation is only useful for theoretical or mathematical analysis. The same goes for the impulse function, as illustrated in Figure 1.29.

They only exist for theoretical derivation and analysis purposes. Therefore, all the variables in the 'if (delta_time >= T) {}' statement are held constant for the entire T sampling period and only change or update in the next iteration. So, the ZOH block is intrinsically present in the void loop() function, it is not a function we implement and it's not something we have to implement directly. Again, the reason we include the sampler and the ZOH block in our diagram is because they mathematically and theoretically represent what is happening to our signals and they are a model representation of what is happening to our signal, theoretically. So, there you have it and that is all you need to implement the following control structure as represented in the above code, and you can also see in Figure 1.30.

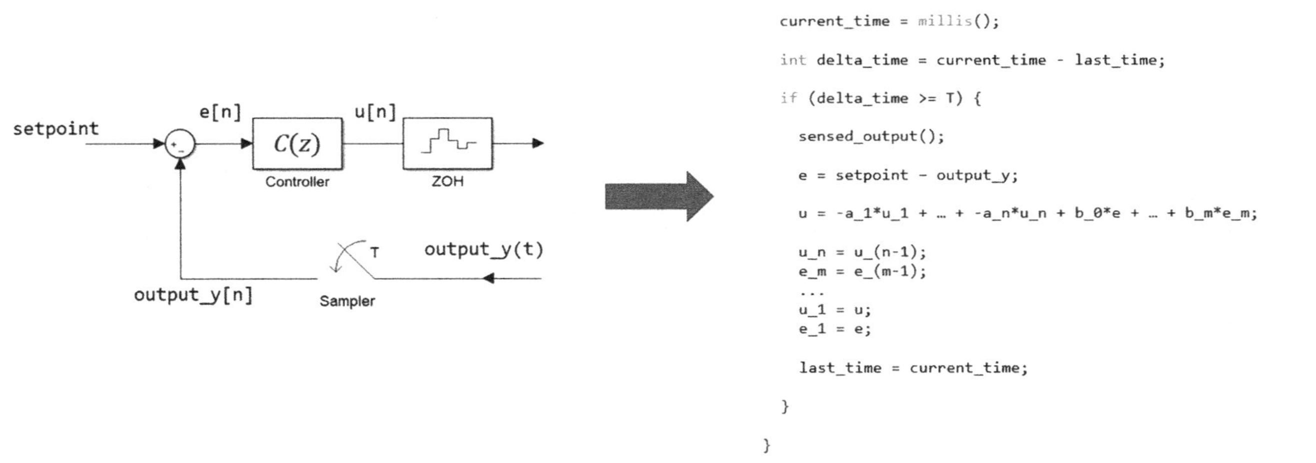

FIGURE 1.30 Arduino code for implementing the control structure block diagram.

So, now that we are in most of the theory out of the way. In the next few sections, we'll dive into the design step process and eventually demonstrate this process using a DC motor speed control example.

1.7 DIGITAL CONTROLLER DESIGN (SIX STEPS)

In this section, we'll be going over the six design steps you can use to design a digital controller to control your continuous-time system, so, here are the six design steps.

1. Obtain the plant's transfer function, $P(s)$
2. Choose an approximate sampling time, T
3. Design controller, $C(s)$, while taking into account the effects of ZOH/sampling
4. Use bilinear transform/Tustin's method to discretize $C(s)$ to $C(z)$
5. Take the inverse Z-transform of $C(z)$ to obtain the difference equation
6. Implement the difference equation on the Arduino and run the system!

We want to let you know that this step-by-step procedure isn't the only way to go about designing a digital controller. This is just a method that makes sense to us and hopefully will make intuitive sense to you. So, the first step is to obtain the continuous-time transfer function of the system you wish to control and we will refer to this transfer function as $P(s)$. Now this step is very open ended because there are many ways to obtain a transfer function of an open-loop system. For example, the method used to obtain the transfer function of a power supply will differ from the method used to obtain the transfer function of a DC motor or a heater system. So, for now, we'll assume you know what the open-loop transfer function is for the system you want to control. There's plenty of literature and papers out there. We'll show you how to get the transfer function of a quad-copter DC motor, heater system, power supply, etc. In the coming section, we'll be demonstrating this step for a DC motor but it is your job to figure out the transfer function of the system you wish to control. So, let's say you have your open-loop transfer function. So, the next step is to choose the sampling frequency or sampling time for your digital controller. So, there is actually no clear-cut method or equation to choose a sampling frequency or sampling time. If you Google this, you'll see that there are several ways to go about choosing a sampling time. However, Tustin's method works better when the sampling frequency ($1/T$) is much higher than the Nyquist frequency. So, what is the Nyquist frequency? Nyquist frequency (f_n) is the minimum sampling frequency needed to prevent introducing aliasing or introducing distortion/error in your closed-loop system. The Nyquist frequency is defined as being strictly twice the bandwidth of your open-loop system as the following equation:

$$f_n \geq 2 \times f_{\text{bw}} \; or \; T_n \leq \frac{T_{\text{bw}}}{2} \tag{1.41}$$

But like we said Tustin's method works better when the sampling frequency is much higher than the Nyquist frequency. So, as a rule of thumb is preferable to use a sampling frequency that's 5–10 times faster than the bandwidth (f_{bw}) of your open-loop system. The bandwidth of your system is typically defined as being the frequency at which the gain of your system is about 3 dB is less than the DC gain. So, the bandwidth frequency or 3 dB frequency is essentially the point where the DC gain begins to roll off, especially at 3 dB below the DC gain. So, in order to find the bandwidth of your system, it is required to know the

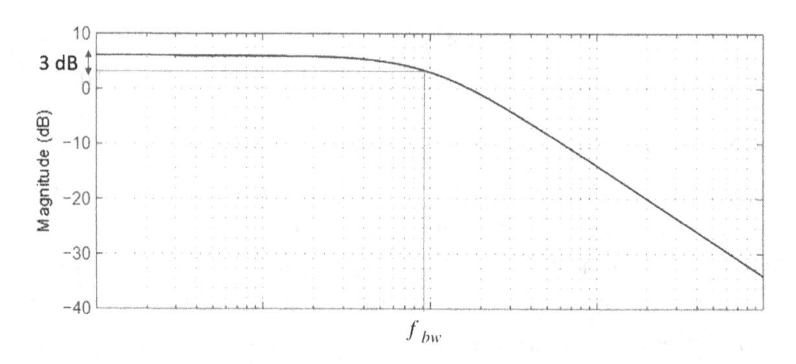

FIGURE 1.31 The bandwidth frequency or 3 dB frequency.

transfer function so that you may plot the Bode plot and locate this minus 3 dB point on the magnitude graph as shown in Figure 1.31.

Next, we move on to design the controller, $C(s)$. So, as we mentioned in the earlier section when needing a digital controller, it's much easier to design it as a continuous mode controller and then discretize it to implement it digitally. So, we got the open-loop transfer function $P(s)$ and we've chosen our sampling frequency so can we start the controller, $C(s)$, design process yet? No, we cannot because we'll be designing a continuous-time controller using continuous-time design methods such as root locus or the frequency response analysis, we need to take into account the continuous-time effects the sampling and the ZOH function will have in the control loop as depicted in Figure 1.32.

If we take a continuous-time signal $x(t)$ and pass it through the sampler and the ZOH block, we get the expected continuous staircase signal which we'll call $x_S(t)$, and if we were to overlap these signals in time you can see the hold delay, and it's discontinuous-time hold delay that we wish to model and include in our control loop before designing our controller as illustrated in Figure 1.33.

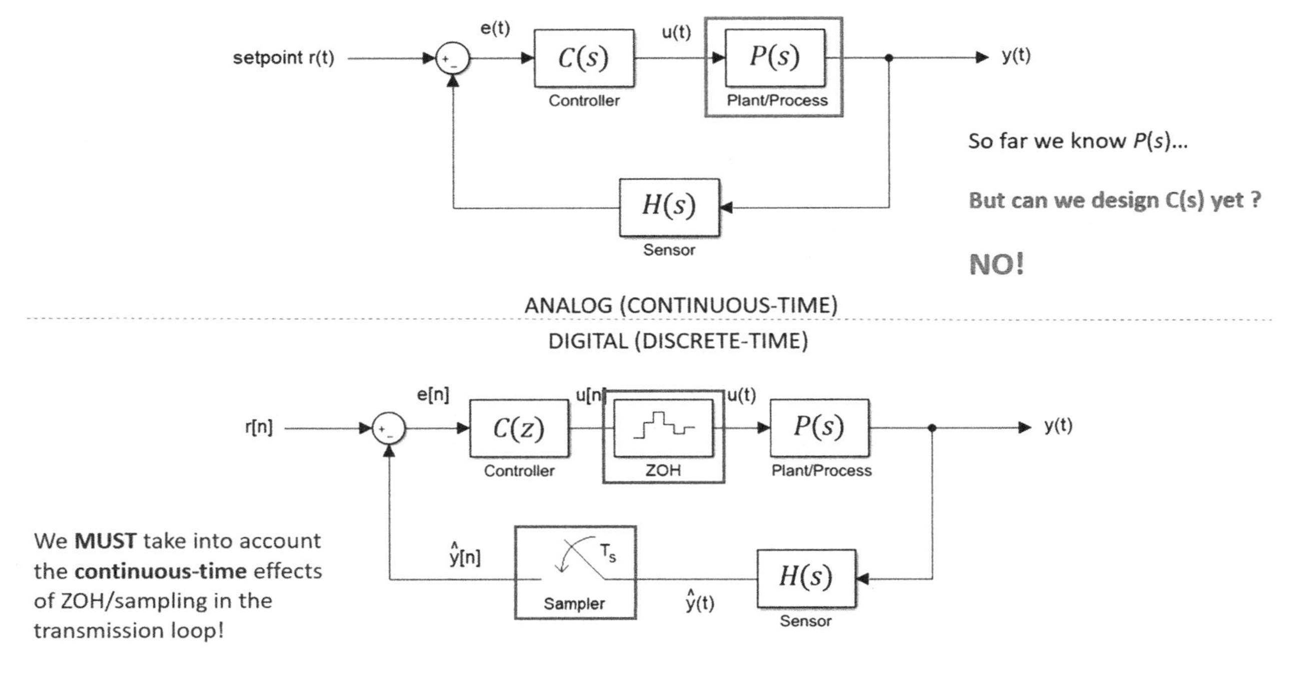

FIGURE 1.32 The continuous-time effects of ZOH/sampling in the transmission loop.

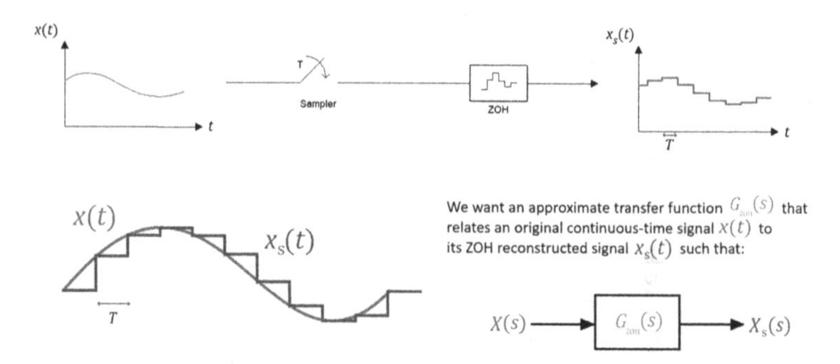

FIGURE 1.33 Approximate transfer function to model the hold delay.

So, we want an approximate transfer function, model this hold delay which we'll call $G_{ZOH}(s)$, that will relate an original continuous-time signal $x(t)$ to its ZOH reconstructed signal $x_s(s)$ as shown in Figure 1.33. So, let's briefly derive $G_{ZOH}(s)$. If we pass a continuous-time signal $x(t)$ through a sampler, we get the following sample signal $x^*(t)$, which we can mathematically represent with the following equation:

$$x^*(t) = x(t)\sum_{k=-\infty}^{+\infty}\delta(t-kT) \tag{1.42}$$

If we take a Laplace transform of $x^*(t)$, we get the following expression:

$$\mathcal{L}\{x^*(t)\} = X^*(s) = \frac{1}{T}\sum_{k=-\infty}^{+\infty}X\left(s - j\frac{2\pi k}{T}\right) \tag{1.43}$$

Now driving this specific equation is beyond the scope of this chapter. So just take our word for it that this is the Laplace transform of $x^*(t)$ as shown in Figure 1.34.

So, if we were to pass $x^*(t)$ through the ZOH block, we get $x_S(t)$. However, $x_S(t)$ can be mathematically represented as a convolution

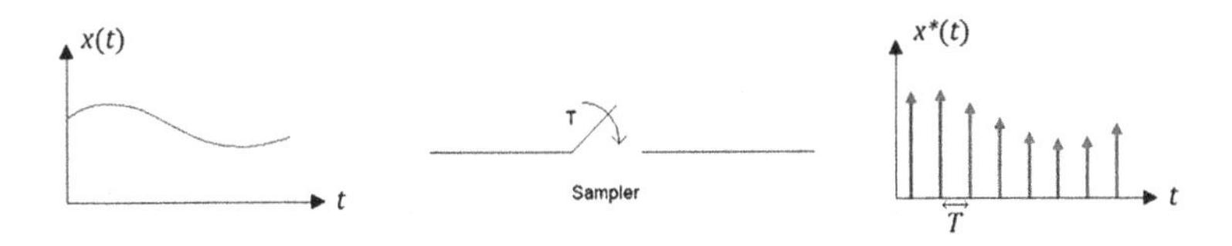

FIGURE 1.34 Brief derivation of $G_{ZOH}(s)$ – Part 1.

between $x^*(t)$ and a function we'll call $g(t)$ where $g(t)$ is the following rectangular function:

$$g(t) = u(t) - u(t-T) \tag{1.44}$$

If we took the Laplace transform of $g(t)$, we get that $G(s)$ equation as below:

$$\mathcal{L}\{g(t)\} = G(s) = \frac{1}{s} - \frac{e^{-sT}}{s} = \frac{1-e^{-sT}}{s} \tag{1.45}$$

Again, T is the sampling period, as depicted in Figure 1.35.

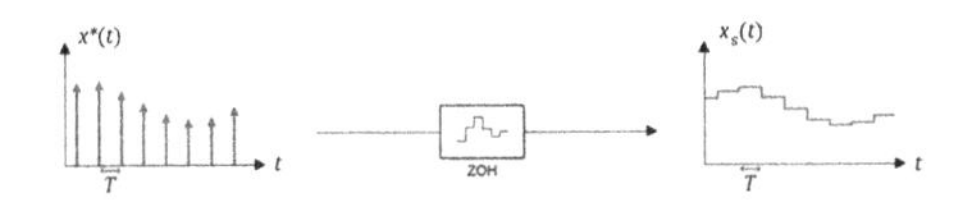

In the time-domain, $x_s(t)$ is obtained by performing convolution between $x^*(t)$ and $g(t)$., where

$g(t) = u(t) - u(t-T)$

$\mathcal{L}\{g(t)\} = G(s) = \frac{1}{s} - \frac{e^{-sT}}{s}$

g(t)

1

T

t

Convolution

$x^*(t) \circledast g(t) = x_s(t)$

$$G(s) = \frac{1-e^{-sT}}{s}$$

FIGURE 1.35 Brief derivation of $G_{ZOH}(s)$ – Part 2.

Now we know from our linear signals and systems fundamentals that the convolution in the time domain translates to multiplication in the frequency domain so we can express $x^{*}(t)$ convolves with $g(t)$ as $X^{*}(s)$ multiplied by $G(s)$ which gets us $x_s(s)$, which is mathematically expressed as the following equations:

$$x^{*}(t) \circledast g(t) = x_s(t) \xrightarrow{\mathcal{L}\{\}} X^{*}(s)G(s) = X_s(s) \tag{1.46}$$

$$\frac{1}{T}\sum_{k=-\infty}^{+\infty} X\left(s - j\frac{2\pi k}{T}\right)G(s) = X_s(s) \tag{1.47}$$

Now of this expression, if we take a closer look at the periodic train function, we only really care about the band-limited component which in itself is simply $X(s)$ or the Laplace transform of our original continuous-time signal $x(t)$. Therefore, realistically we only care about band-limited component ($k = 0$), which is simply $X(s)$. So, we can equate the periodic train function to just $X(s)$ as illustrated in Figure 1.36.

Thus, our expression has been simplified to the following equation:

$$\frac{1}{T}X(s)G(s) = X_s(s) \tag{1.48}$$

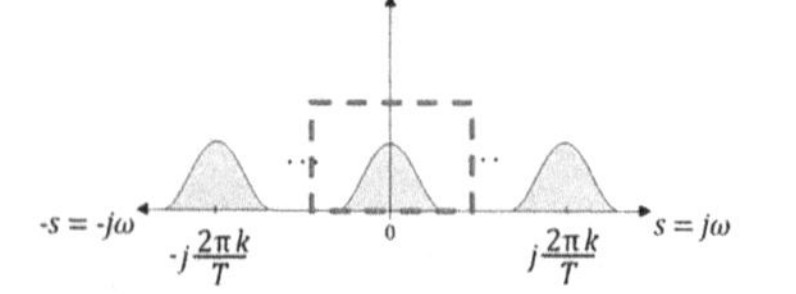

However, realistically we ONLY care about **bandlimited component (*k* = 0)**, which is simply ***X*(*s*)**...

FIGURE 1.36 Brief derivation of $G_{ZOH}(s)$ – Part 3.

We know that $G(s)$ equals to Equation (1.45). So, substituting this we get the following transfer function that relates the staircase function $x_S(t)$ to our original continuous-time signal $x(t)$:

$$(1.39),(1.36) \rightarrow \frac{X_s(s)}{X(s)} \approx \frac{1-e^{-sT}}{sT} \tag{1.49}$$

So, $G_{ZOH}(s)$ is approximately equal to the following equation:

$$G_{ZOH}(s) = \frac{1-e^{-sT}}{sT} \tag{1.50}$$

So, this is the transfer function block that models the hold delay in the continuous-time domain as shown in Figure 1.37.

Now, we'll take this block and we'll place it in the closed-loop, and now we can finally go ahead and design $C(s)$ while taking into account the continuous-time effects of the discrete parts. Therefore, we can now design $C(s)$ using classical design techniques because we accounted for the discrete effects in our control loop as depicted in Figure 1.38. The closed-loop transfer function now looks like the following equation:

$$\frac{Y(s)}{R(s)} = \frac{C(s)P(s)}{1+C(s)P(s)G_{ZOH}(s)H(s)} \tag{1.51}$$

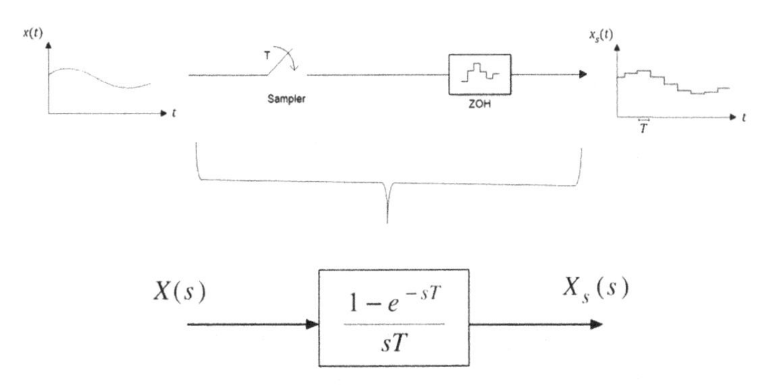

FIGURE 1.37 Modeling the hold delay in the continuous-time domain.

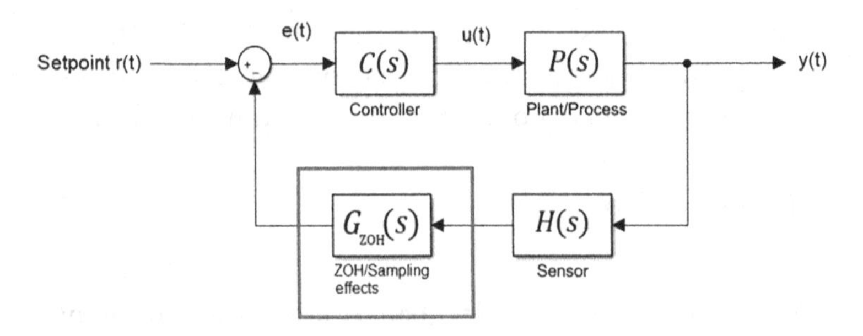

FIGURE 1.38 Adding the hold delay block in the closed-loop.

Now, after designing the continuous-time controller *C*(*s*), we want to discretize it using Tustin's method (bilinear transformation). So, knowing that the sampling time is *T* and knowing that we're going to use Tustin's method we get the following z-domain transfer function:

$$C(s)\Big|_{\substack{s=\frac{2}{T}\frac{z-1}{z+1} \\ T=\text{chosen sampling time}}} = C(z) \tag{1.52}$$

$$C(z)=\frac{U(z)}{E(z)}=\frac{b_0+z^{-1}b_1+z^{-2}b_2+\ldots+z^{-M}b_M}{a_0+z^{-1}a_1+z^{-2}a_2+\ldots+z^{-M}a_M} \tag{1.53}$$

From our transfer function equation, (1.53), we cross multiply the terms as below:

$$\begin{aligned} &U(z)\left(a_0+z^{-1}a_1+z^{-2}a_2+\ldots+z^{-M}a_M\right) \\ &\quad =E(z)\left(b_0+z^{-1}b_1+z^{-2}b_2+\ldots+z^{-M}b_M\right) \end{aligned} \tag{1.54}$$

Next, we take the inverse Z-transform aka going to discrete-time domain, doing so; we end up with a different equation of our controller that relates our error input to the controller output as the following expression:

$$a_0u[n]+a_1u[n-1]+a_2u[n-2]+\ldots+a_Mu[n-M]$$
$$=b_0e[n]+b_1e[n-1]+b_2e[n-2]+\ldots+b_Me[n-M] \quad (1.55)$$

Once we've got that we can go straight to implementing it on the Arduino, and this is the code for the feedback control structure that we discussed in the previous section, and that is it. Those are all the preliminary steps needed to design and implement a simple digital controller for your system. In the last few sections, will apply these six steps to design a digital controller for a DC motor speed control system.

1.8 DC MOTOR SPEED CONTROL

1.8.1 System Modeling

In this section, we're going to dive into designing a digital feedback control system to regulate the speed of a real DC gearmotor. The block diagram of the control loop is shown in Figure 1.39.

In this case, we will have the PWM motor driver block after the ZOH block and the plant $P(s)$ is the DC gearmotor, $H(s)$ is the encoder, and output is the velocity ($\dot{\theta}(t)$) in rad/s. We want to implement the rotary encoder that is used to acquire the motor speed which then gets sampled by the Arduino as a discrete signal which we'll call output speed. We then compare it with a desired speed variable to generate an error signal, *e*. This error signal then goes through our controller $C(z)$ which again is implemented as a

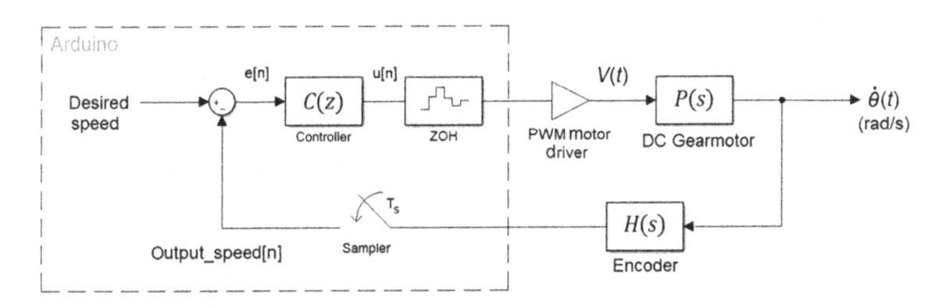

FIGURE 1.39 The block diagram of the control loop.

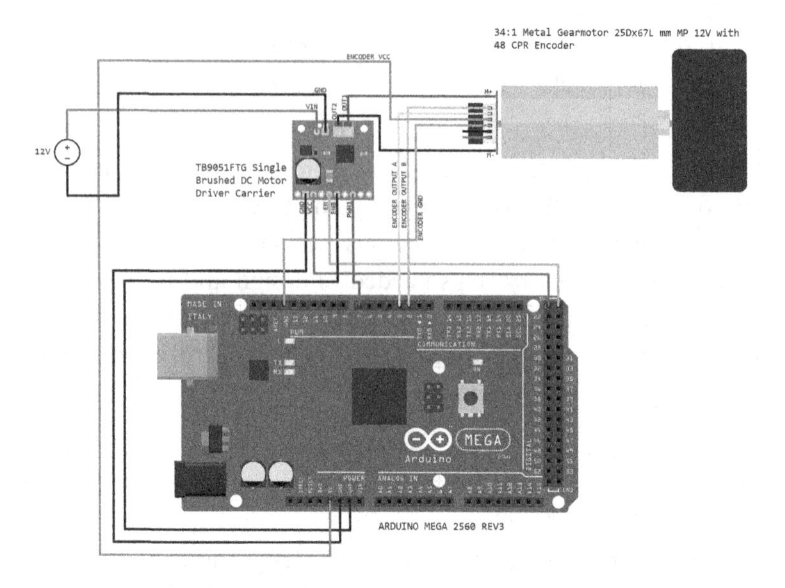

FIGURE 1.40 The schematic of the setup.

difference equation to generate a control signal. This control signal is realized as a PWM signal that comes out of the Arduino and goes into the PWM motor driver. The motor driver imposes an input voltage on the gearmotor to bring the actual speed, $\dot{\theta}(t)$, toward desired speed and thus we have our feedback loop. Here in Figure 1.40 is the apparatus for the DC motor control system. We have our Arduino Mega 2560 which is acting as our digital controller followed by a PWM motor driver which drives our gearmotor. Our motor conveniently has a built-in encoder that sends pulses data to where do we know which contains an algorithm to convert this data to an output speed value. The motor driver receives this power from the DC power supply and the real output voltage is set to 12 volts. The schematic of the setup is depicted in Figure 1.40.

So, we want to design a digital controller to effectively regulate the speed of this DC gearmotor. So, the following are the six-step procedures as discussed in the previous section.

1. Obtain the plant's transfer function $P(s)$
2. Choose an approximate sampling time T
3. Design controller $C(s)$ while taking into account the effects of ZOH/sampling
4. Use bilinear transform/Tustin's method to discretize $C(s)$ to $C(z)$
5. Take the inverse Z-transform of $C(z)$ to obtain the difference equation
6. Implement the difference equation on the Arduino and run the system!

Our first task is to obtain the approximate transfer function of our DC gearmotor. So, in order to do that we must drive a transfer function based on the electro-mechanical system model of a generic DC motor as illustrated in Figure 1.41.

Therefore, in order to design a speed control system for this gearmotor using classical control techniques, we must first derive its transfer function, $P(s)$, using an equivalent system model that approximately models its dynamics. Now running out all the physics for the mathematical model derivation is beyond the scope of this chapter. So, we'll go straight into the transfer function of

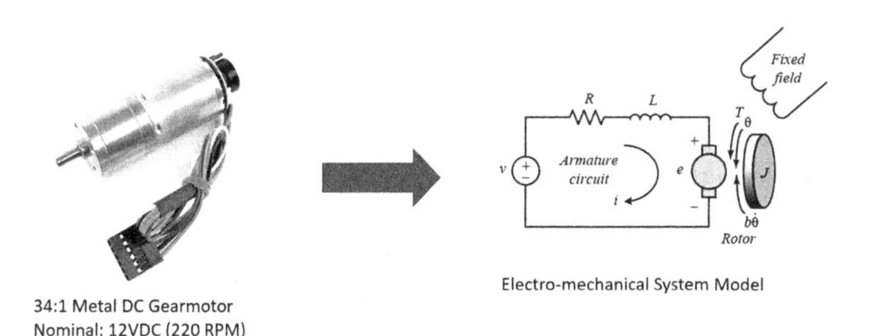

FIGURE 1.41 The DC motor transfer function – Part 1.

this model that relates the input voltage, $V(s)$, to the output speed, $\dot{\theta}(s)$, as the following equation [4]:

$$P(s)=\frac{\dot{\theta}(s)}{V(s)}=\frac{s\theta(s)}{V(s)}=\frac{K}{(Js+b)(Ls+R)+K^2}\left[\frac{\text{rad/s}}{\text{V}}\right] \tag{1.56}$$

where J (kg.m^2) is the moment of inertia of the rotor, b (N.m.s) is the motor viscous friction constant, K is equal to the electromotive force constant (K_e(V/rad/s)) or motor torque constant (K_t(N.m/Amp)), R(ohm) is the electric resistance, and $L(H)$ is the electric inductance. From Figure 1.41, we can derive the following governing equations based on Newton's second law and Kirchhoff's voltage law:

$$J\ddot{\theta}+b\dot{\theta}=K_t i=Ki \tag{1.57}$$

$$L\frac{di}{dt}+Ri=V-K_e\dot{\theta}=V-K\dot{\theta} \tag{1.58}$$

By applying the Laplace transform to Equations (1.57) and (1.58), we will have the following equations:

$$(Js+b)s\theta(s)=KI(s) \tag{1.59}$$

$$(Ls+R)I(s)=V(s)-Ks\theta(s) \tag{1.60}$$

By eliminating $I(s)$ from Equations (1.59) and (1.60), we get the open-loop transfer function in Equation (1.56). Now, if you wish to see where we got this transfer function from, you can find the detailed derivation in the link provided in [4]. So, the five parameters of the DC motor transfer function are as below as shown in Figure 1.42.

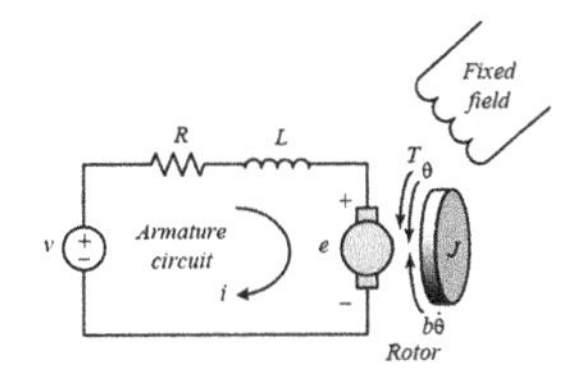

Electro-mechanical System Model

J = rotor's moment of inertia (kg.m^2)

b = motor's viscous friction constant (N.m.s)

K = electromotive force constant (V/rad/s)

R = electric resistance (Ω)

L = electric inductance (H)

$$P(s) = \frac{\dot{\Theta}(s)}{V(s)} = \frac{K}{(Js+b)(Ls+R)+K^2}$$

FIGURE 1.42 The DC motor transfer function – Part 2.

J is the rotor's moment of inertia (kg.m^2)

b is the motor's viscous friction constant (N.m.s)

K is the electromotive force or motor torque constant (V/rad/s)

R is the electric resistance (Ω)

L is the electric inductance (H)

So, now the big question is how do we find these values for our gearmotor? Well, if you're lucky all this information can be available on the motor's data sheet. However, almost 99% of the time this is never the case, so, we'll have to somehow figure out these values using our tools and our intuition. So, let's try to figure out K, now K is the electric motor force or torque constant. This is essentially the DC gain that relates our input voltage to our output speed in radians per second (rad/s). Now, according to the data sheet, the nominal speed at the rated voltage of 12 volts is 220 RPM now, 220 RPM can also be expressed as 23 rad/s since we have:

$$220\,\text{RPM} = 220 \times \frac{2\pi(\text{rad})}{60(\text{s})} \approx 23\,\text{rad/s} \tag{1.61}$$

So, dividing 12 V by 23 rad/s, we get an approximate constant of about 0.52 V/rad/s. Now what about the resistance? Now luckily, we have an LCR meter which can measure electrical resistance of complex impedance. So, if we probe across the leads of the motor, we can get an approximate value of the internal resistance. Now the resistance and the inductance actually vary with the angle of the rotor. But again, we're just looking for an approximate value. So, for now, we'll just say that the resistance is around 4.4 ohms. Now using the same meter which is capable of measuring inductance values at several frequencies we measure the inductance to be around 6.3 mH, at an applied low frequency of 100 Hertz. So far, we've gotten three values down, great, but now how do we go about finding J and b. Now, unless we have some very expensive precision tests' apparatus, finding even the approximate values for these parameters is quite difficult. Luckily, MATLAB Simulink has a parameter estimation feature in the optimization toolbox. The design optimization toolbox comes with a parameter estimation tool, where the user can estimate the parameters of a simulated model of their system using real measured input and output data.

So, the first step to use a tool is to create a Simulink model based on the differential equations that describe the dynamics of the system or to create a linear model that captures the dynamics of a DC motor (Equation 1.56 or Equations 1.59 and 1.60) as shown in Figure 1.43.

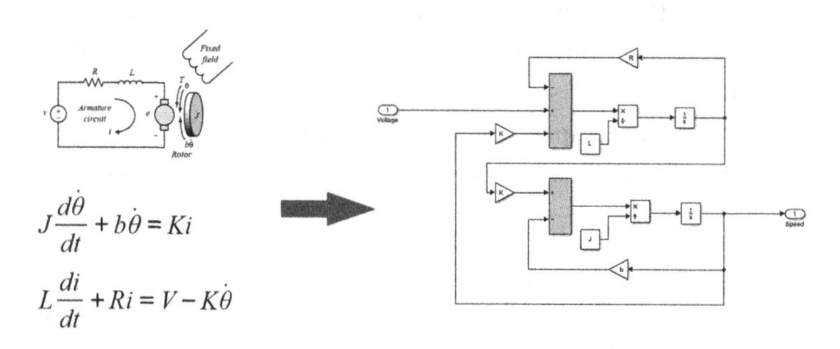

FIGURE 1.43 The Simulink model based on dynamic equations.

Now, if you've never played with Simulink or modeled on Simulink, we highly suggest that you get exposed to it because it's very useful for control design and simulating real-time systems. There are plenty of tutorials all over YouTube and Udemy on modeling using MATLAB Simulink, so we won't be discussing that here as it's beyond what this chapter focuses on. The second step is to obtain the input and output data we want to use. So, in this Arduino script, we've created a test input signal that will vary the input voltage applied to the motor in a step-like fashion and will want to monitor the elapsed time, the applied input and the output speed.

```
int sampling_time = 50;
unsigned long previous_time, current_time;
float encoder_count = 0;
boolean encoder_A, encoder_B;
byte state, statep;
float previous_angle, current_angle = 0;
float motor_speed = 0;
float supply_voltage = 12;
float elapsed_time = 0;
const int pwm_output = 7;
float control;
void Encoder_State(){
 encoder_A = digitalRead(3);
 encoder_B = digitalRead(2);
 if ((encoder_A == HIGH) && (encoder_B == HIGH))
 state = 1;
 if ((encoder_A == HIGH) && (encoder_B == LOW))
 state = 2;
 if ((encoder_A == LOW) && (encoder_B == LOW))
 state = 3;
 if ((encoder_A == LOW) && (encoder_B == HIGH))
 state = 4;
 switch (state)
 {
  case 1:
```

```
  {
   if (statep == 2) encoder_count--;
   if (statep == 4) encoder_count++;
   break;
  }
   case 2:
  {
   if (statep == 1) encoder_count++;
   if (statep == 3) encoder_count--;
   break;
  }
   case 3:
  {
   if (statep == 2) encoder_count++;
   if (statep == 4) encoder_count--;
   break;
  }
  default:
  {
   if (statep == 1) encoder_count--;
   if (statep == 3) encoder_count++;
   break;
  }
 }
 statep = state;
}
void get_speed(){
 current_angle = (encoder_count*360.0)/(1632.67);
 motor_speed = 0.01745*((current_angle
 - previous_angle)/0.05);
 previous_angle = current_angle;
}
void step_input(){
 control = 0;
 if (current_time >= 2000 && current_time <=
 6000){
   control = 12;
  }
```

```
  if (current_time > 6000 && current_time <=
  10000){
    control = 6;
   }
  if (current_time > 10000 && current_time <=
  14000){
    control = 8;
   }
  if (current_time > 14000){
    control = 0;
  }
}
void setup(){
 Serial.begin(9600);
 previous_time = millis();
 pinMode(2,INPUT);
 pinMode(3,INPUT);
 attachInterrupt(digitalPinToInterrupt(2),
 Encoder_State, CHANGE);
 attachInterrupt(digitalPinToInterrupt(3),
 Encoder_State, CHANGE);
 encoder_A = digitalRead(3);
 encoder_B = digitalRead(2);
 if ((encoder_A == HIGH) && (encoder_B == HIGH))
 statep = 1;
 if ((encoder_A == HIGH) && (encoder_B == LOW))
 statep = 2;
 if ((encoder_A == LOW) && (encoder_B == LOW))
 statep = 3;
 if ((encoder_A == LOW) && (encoder_B == HIGH))
 statep = 4;
 pinMode(pwm_output, OUTPUT);
}
void loop() {
current_time = millis();
int delta_time = current_time - previous_time;
if (delta_time > sampling_time){
 elapsed_time = current_time/1000.0;
```

```
step_input();
get_speed();
analogWrite(pwm_output,(control/
supply_voltage)*255.0);
Serial.print(control);
Serial.print("    ");
Serial.print(elapsed_time);
Serial.print("    ");
Serial.println(motor_speed);
previous_time = current_time;
 }
}
```

Let's run the script and then from Tools menu, we select the Serial Monitor option. Now we'll take 15 seconds rows of data and then on MATLAB, we'll create 7 data variables that contain *R*, *L*, *K*, *J*, *b*, and also, the measured Input_Data and Output_Data which the first column is the elapsed_time and the second column is control and motor_speed, respectively, as depicted in Figure 1.44.

Now, before the estimation tool can run, we'll need to use some initial values for all the parameters. So earlier we approximated what *K*, *R*, and *L* could be. So, that's what we'll be using for their initial values as illustrated in Figure 1.45.

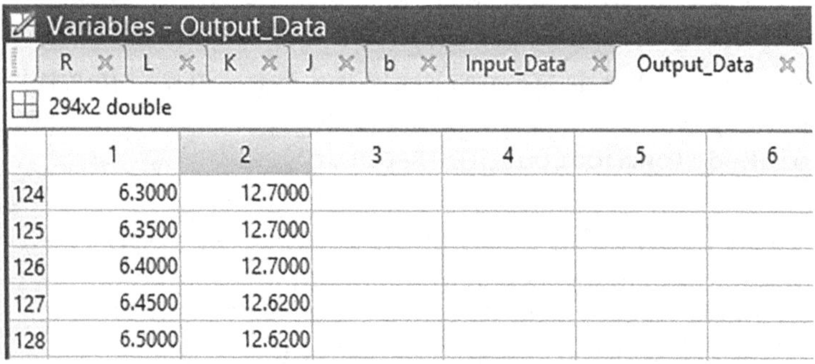

Variables - Output_Data

R | L | K | J | b | Input_Data | Output_Data

294x2 double

	1	2	3	4	5	6
124	6.3000	12.7000				
125	6.3500	12.7000				
126	6.4000	12.7000				
127	6.4500	12.6200				
128	6.5000	12.6200				

FIGURE 1.44 Creating variables that contain the measured data.

$J = ?$

$b = ?$

$K = v(t)/\theta(t) = 12\text{ V} / 23\text{ rad/s} = 0.52\text{ V/rad/s}$

$R = 4.40\ \Omega$

$L = 6.3\text{ mH}$

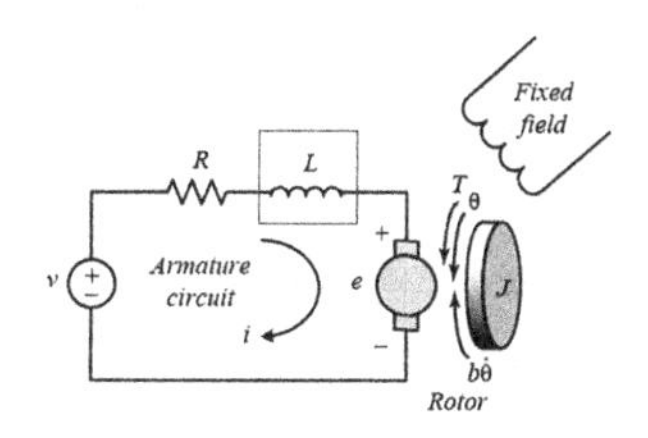

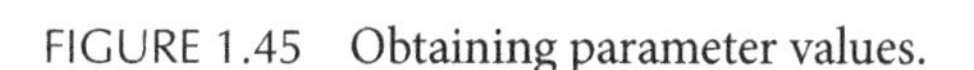

FIGURE 1.45 Obtaining parameter values.

However, for b and J, we'll just pick some random values to begin with (J = 0.001, b = 0.1). Now, we are opening up the Simulink, and the file containing our motor model.

We'll go to the apps tab and then open up parameter estimator. From this window, we want to create a New Experiment. Next, we want to pull the measured Output_Data and Input_Data from the workspace and then we want to select the parameters (Select Parameters), since we wish to estimate all those five parameters (J, K, L, R, and b). Therefore, we select all of them and click on the OK button.

Now, in order to get more realistic parameter estimation, we want to set the Min and Max boundaries so that the tool doesn't vary any parameters too far from a realistic value. So, we'll set the Min and Max for K, R, and L to what we think they could be, and because we don't know b and J, we'll leave its max alone but we'll want to set the minimum value to zero because realistically none of these parameters should ever go negative as below.

J = 0.001, Minimum = 0, Maximum = Inf

K = 0.52, Minimum = 0.47, Maximum = 0.54

L = 0.0063, Minimum = 0.006, Maximum = 0.007

R = 4.4, Minimum = 3.8, Maximum = 5

b = 0.1, Minimum = 0, Maximum = Inf

Now, if you click on Plot button in that window, you should see the measured input voltage and output speed data we got from the Arduino.

Now let's click on the Estimate button, you can see that the tool is running some sort of optimization algorithm to curve fit the simulation output to the measured output and it's doing this by iterating and varying the parameters. Now, once we've come to a close enough fit, the tool stops estimating and our parameter values should have automatically been updated by clicking on the Exp icon in Experiments sub-window.

Now, using the following script as given below, we can get the approximate transfer function for our DC gearmotor as the following equation:

$$P(s) = \frac{0.47}{3.425e-5s^2 + 0.01859s + 0.2211} \tag{1.62}$$

```
J = 0.004893;
b = 5.8147e-5;
K = 0.47;
R = 3.8;
L = 0.006999;
num = K;
den = [J*L, (J*R)+(b*L), (b*R)+(K^2)];
sys = tf(num,den);
```

So, step 1, which is obtaining the plant's transfer function *P*(*s*), is complete. So, let's stop here for now, until now, we've gotten an approximate transfer function. In the next section, we'll move on to selecting an appropriate sampling time and design your controller.

1.8.2 Controller Design

In this section, we will continue the controller design process for DC motor speed control example. So, we dedicated the previous section to obtaining an approximate transfer function for DC gearmotor which is shown in Figure 1.46.

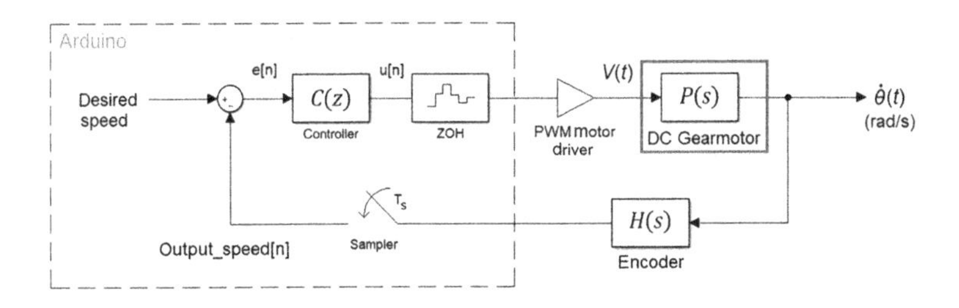

$$P(s) = \frac{0.47}{3.425e-05s^2 + 0.01859s + 0.2211}$$

FIGURE 1.46 Obtaining an approximate transfer function for DC gearmotor.

So, moving on to step 2, we must find a sampling time, T, for our control loop. So, as we discussed a couple of sections ago, the sampling frequency should be at least 5–10 times higher than the bandwidth of the open-loop system, $P(s)$. Using the following MATLAB code, we can obtain the Bode plot for our system.

```
s = tf('s');
P_s = 0.47/(3.425e-5*s^2 + 0.01859*s + 0.2211);
bode(P_s);
grid on;
```

From observing the magnitude plot, we can see that the DC gain is around 6.55 dB and the bandwidth of the system is at the frequency 3 dB below the DC gain, so, moving down 3 dB below, we get to 3.55 dB and the frequency at this gain 12.1 rad/s or 1.93 Hz. So, the bandwidth of our system is 1.93 Hz or around 2 Hz. So, we'll multiply this by 10 and arrive at a sampling frequency of 20 Hz which correlates to a sampling time of 50 milliseconds as shown in Figure 1.47 and equation below:

$$T = \frac{1}{f} = \frac{1}{20} = 0.05\,\text{seconds} \tag{1.63}$$

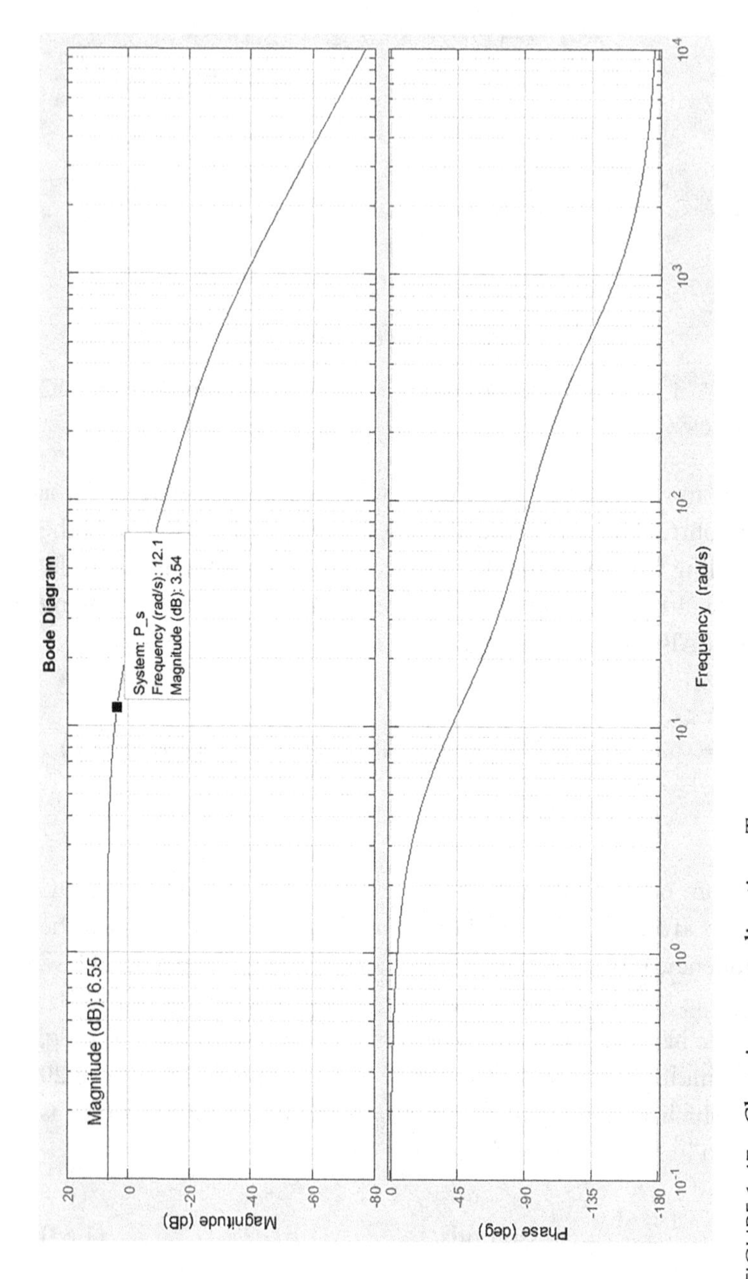

FIGURE 1.47 Choosing a sampling time, T.

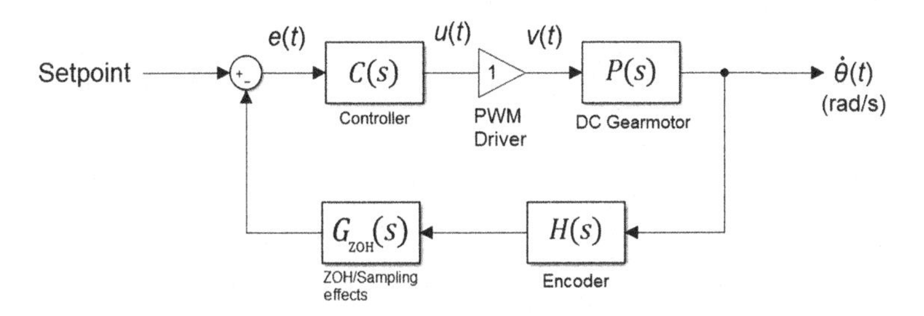

FIGURE 1.48 The continuous-time block diagram for our control loop.

So now that, we have our sampling time and step 2 is complete. Now let's move on to designing a continuous-time controller *C*(*s*), as shown in Figure 1.48, which is the continuous-time block diagram for our control loop and here after *C*(*s*) we have a PWM driver block with gain of 1, the *P*(*s*) is the transfer function of DC gearmotor obtained in the previous section, and *H*(*s*) is the transfer function of encoder with the gain of 1. Because the encoder is simply providing pulse count data that is to be interpreted by the Arduino as a speed value in rad/s, we can just set the encoders transfer function, *H*(*s*), as a gain of 1, so we have the following equation:

$$H(s)=1 \tag{1.64}$$

So, this is our closed-loop transfer function that relates a set point to our output speed as the following equation:

$$G_{\dot{\theta}R}(s)=\frac{\dot{\theta}(s)}{R(s)}=\frac{C(s)P(s)}{1+C(s)P(s)G_{\text{ZOH}}(s)H(s)} \tag{1.65}$$

Also, knowing that each *H*(*s*) is equal to 1, Equation (1.54), we get the following expression:

$$G_{\dot{\theta}R}(s)=\frac{\dot{\theta}(s)}{R(s)}=\frac{C(s)P(s)}{1+C(s)P(s)G_{\text{ZOH}}(s)}=\frac{C(s)P(s)}{1+T(s)} \tag{1.66}$$

$$T(s)=C(s)P(s)G_{\text{ZOH}}(s)H(s) \tag{1.67}$$

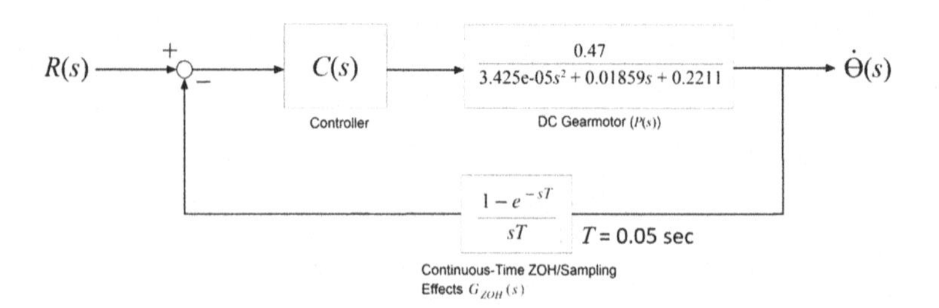

FIGURE 1.49 Adding explicit transfer functions for $P(s)$ and $G_{\text{ZOH}}(s)$.

With $T(s)$, is being our loop gain. Here as depicted in Figure 1.49 is a closed-loop block diagram except with the explicit transfer functions shown for $P(s)$ and $G_{\text{ZOH}}(s)$.

Therefore, the explicit transfer functions for $G_{\text{ZOH}}(s)$ are given in the following equation:

$$G_{\text{ZOH}}(s) = \frac{1-e^{-sT}}{sT}, \quad T = 0.05\,\text{sec} \tag{1.68}$$

Now, it's quite apparent that designing a controller from this point can be quite difficult because $G_{\text{ZOH}}(s)$ has an exponential term and therefore it's irrational. We should approximate $G_{\text{ZOH}}(s)$ further to a rational function to make the control design easier for MATLAB. The control design tools in MATLAB tend to work more nicely with the rational transfer functions as opposed to transfer functions that have exponential terms in them. So, using an approximation technique known as the Pade approximation, the exponential term can be approximated as a rational function with ever-increasing orders of precision, so the higher the order, the better the approximation as the equation below:

$$\text{Pade Appr.}: e^{x} = \frac{1+\left(\frac{1}{2}\right)x+\left(\frac{1}{9}\right)x^{2}+\left(\frac{1}{72}\right)x^{3}+\left(\frac{1}{1008}\right)x^{4}+\left(\frac{1}{30240}\right)x^{5}\ldots}{1-\left(\frac{1}{2}\right)x+\left(\frac{1}{9}\right)x^{2}-\left(\frac{1}{72}\right)x^{3}+\left(\frac{1}{1008}\right)x^{4}-\left(\frac{1}{30240}\right)x^{5}\ldots} \tag{1.69}$$

Now, of course, doing this by hand is impractical but luckily MATLAB has a function (available with Control System Toolbox) called pade(sys,n) with sys = transfer function, n = approximation order, which if you give it the irrational transfer function and a desired order n, it will output the approximate rational transfer function. So, using the following MATLAB script, we can obtain the following rational transfer function for $G_{ZOH}(s)$ and we'll call that $G_{ZOHP}(s)$:

$$G_{ZOHP}(s) = \frac{40s^3 + 960000s}{s^4 + 240s^3 + 24000s^2 + 960000s} \tag{1.70}$$

```
T = 0.05;    %sampling time T = 0.05 seconds
s = tf('s');
G_zoh = (1-exp(-s*T))/(s*T);  %original ZOH/
sampling delay transfer function
G_zohp = pade(G_zoh,3);  % 3rd order approximation
```

So now that, we know that both $P(s)$ and $G_{ZOHP}(s)$ are rational and we can go ahead and design $C(s)$ as shown in Figure 1.50.

Now, for controlling the speed and position of a DC motor, typically a proportional integral (PI) controller is used in the industry and most conventional DC motors have their output speed regulated using a PI controller as the following equation:

$$C(s) = K_p + \frac{K_i}{s} \tag{1.71}$$

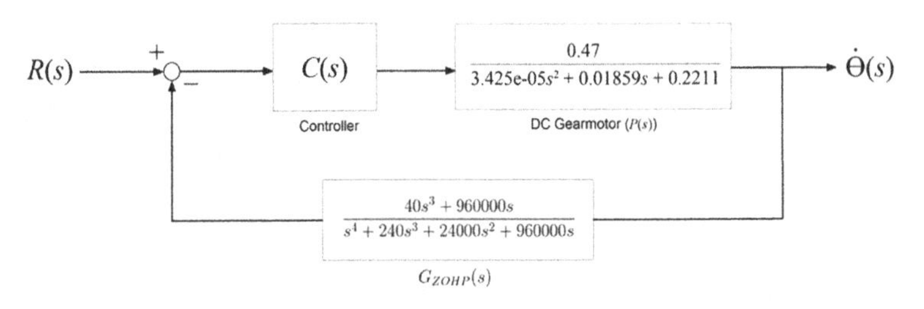

FIGURE 1.50 The rational transfer functions for $P(s)$ and $G_{ZOHP}(s)$ in the control loop.

So now, we can design *C*(*s*) using classical control techniques such as root locus, frequency response analysis, or some other complicated control design method but look at the transfer function we're dealing with in total they make up a six-order system. Now if you want to do root locus or frequency response analysis on a six-order system, you can. However, we think it's a lot more convenient to use the control design tools that are available with MATLAB. Now luckily because we're designing a controller that's a part of the proportional integral derivative (PID) family, we can use the pidTuner(sys) app that's available with the control system toolbox and the code is as below.

```
T = 0.05;    %sampling time T = 0.05 seconds
s = tf('s');
P_s = 0.47/(3.425e-5*s^2 + 0.01859*s + 0.2211);
%DC gear motor's transfer %function P(s)
G_zoh_s = (1-exp(-s*T))/(s*T);  %ZOH/Sampling's
continuous-time effect transfer function G_zoh(s)
G_zohp_s = pade(G_zoh_s,3);  %3rd order Pade
approximation of G_zoh(s)
sys = P_s*G_zohp_s;  %total open-loop gain
pidTuner(sys);  %launch PID Tuner App on MATLAB
```

We declare our transfer functions *P*(*s*) and G_{ZOHP}(s) and then we multiply them to get the total effect of open-loop transfer function. Then we store it in a variable which we'll call 'sys' which is short for system. We then call the pidTuner(sys) function with 'sys' being the input argument. Now, before we run the script, we want to set some design requirements. We would like the settling time for the output transient to be less than 2 seconds. The overshoot and undershoot to be under 5% of the steady-state value and the steady-state error to be less than 3% of the steady-state value. So, the design requirements are:

- Settling time less than 2 seconds
- Overshoots and undershoots should be under 5%
- Steady-state error should be less than 3%

So now let's run the script, so the PID tuner raptured open up automatically and by default it seems like it chose a PI controller on its own, and what you see now is a step response of the closed-loop system's output and upon the top that are sliders that we can use to adjust the response time and transition behavior such that it meets our criteria, and on the bottom right should be the K_p and the K_i gain values that it correlates to. So, we'll play around with this plot until we think it looks good. So, this response seems to work and if we click on the Show Parameters tab, we can see the gain values, the settling time, overshoot percentage, and the stability margins as depicted in Figure 1.51.

So far, they all meet our criteria and we think we'll go with this design, so, this is our continuous-time PI controller with K_p that is equal to 0.427 and K_i is equal to 5.114 and we have the following equation:

$$C(s) = K_p + \frac{K_i}{s} = 0.427 + \frac{5.114}{s} \tag{1.72}$$

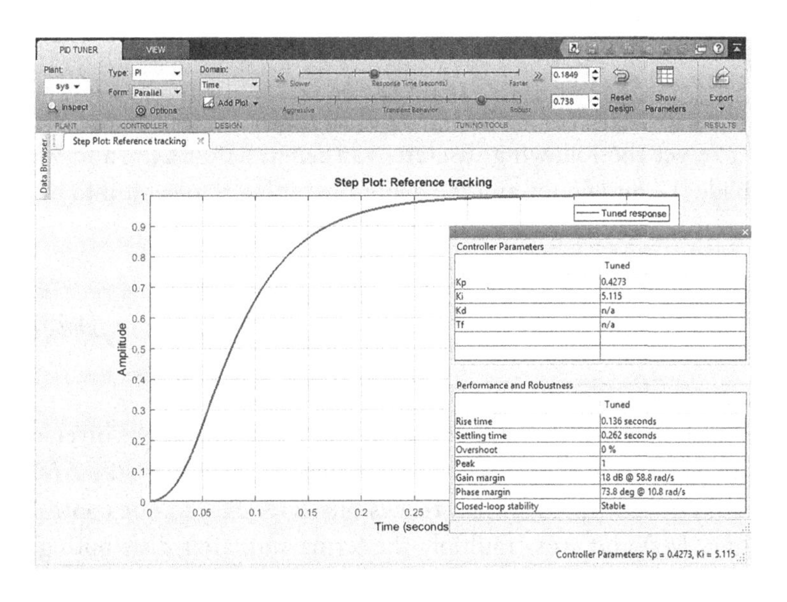

FIGURE 1.51 Clicking on the Show Parameters tab.

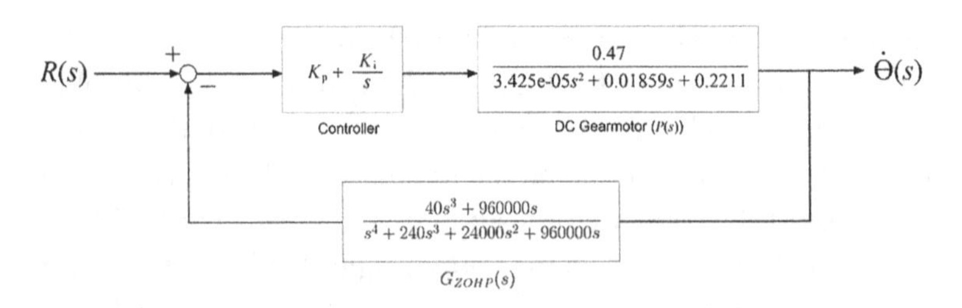

FIGURE 1.52 Complete design of continuous-time control loop.

So now that, we have got our continuous-time controller, and step 3 is complete as illustrated in Figure 1.52.

Now, we have to discretize it using Tustin's method by running the following MATLAB script that contains a c2d function with the method set to Tustin.

```
T = 0.05;   %sampling time T = 0.05 seconds
s = tf('s');
C_s = 0.427 + (5.114/s);  %continuous-time
controller C(s)
C_z = c2d(C_s,T,'tustin');  %discretized
controller C(z)
```

We get the following discrete transfer function $C(z)$, and we divide the numerator and the denominator by 'z' to get it into his causal expression as the equation below:

$$C(z)=\frac{U(z)}{E(z)}=\frac{0.5548z-0.2991}{z-1}=\frac{0.5548-0.2991z^{-1}}{1-z^{-1}} \tag{1.73}$$

So now, step 4 is complete and now we have to take the inverse Z-transform to get the difference equation. Now we take $C(z)$ from Equation (1.73) which relates our error $E(z)$ to our control signal $U(z)$ and cross multiply the terms and, after distributing, we then take the inverse Z-transform to get a difference equation as the following equations:

$$U(z)\left(1-z^{-1}\right)=E(z)\left(0.5548-0.2991z^{-1}\right) \tag{1.74}$$

$$U(z)-U(z)z^{-1}=E(z)0.5548-E(z)0.2991z^{-1} \tag{1.75}$$

$$(1.64)\xrightarrow{Z^{-1}\{\}}u[n]-u[n-1]=0.5548e[n]-0.2991e[n-1] \tag{1.76}$$

Then we bring all the delayed input and output terms to the right side as the equation below:

$$u[n]=u[n-1]+0.5548e[n]-0.2991e[n-1] \tag{1.77}$$

Now, we've obtained the controller's difference equation that relates our discrete error signal $e[n]$ to our discrete control output $u[n]$. So now, step 5 is complete. Now, all we have to do is to implement the controller on the Arduino and see the closed-loop speed control in action and we'll do this in the next section.

1.8.3 Arduino Demonstration

In this final section, we'll finally implement the digital controller designed in the previous section to regulate the speed of our DC gearmotor. So, we dedicated the last two sections to completing steps 1–5. Now all that remains is implementing the controller on the Arduino to demonstrate the closed-loop control. However, before we go ahead and code up the script, we are going to first simulate this control structure to see what type of behavior we should expect. Shown here in Figure 1.53 is the Simulink block diagram for our closed-loop feedback system. Notice that there's a saturation block after the controller. Now what does this block will do is to limit the control signal to be within the bounds of

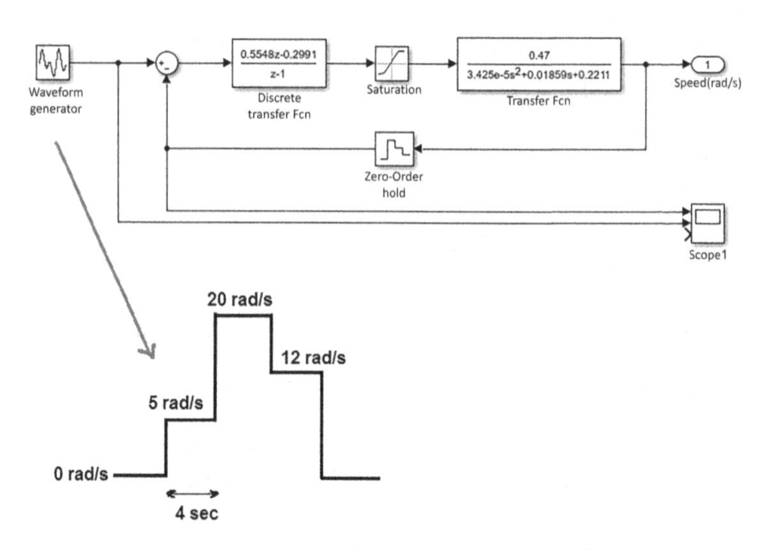

FIGURE 1.53 The Simulink block diagram for our closed-loop feedback system.

0–12 volts and that's because our real power supply for our PWM driver is set to 12 volts.

Also, because the control signal of 12 V correlates to a PWM duty cycle of 100% so it's inherently the maximum voltage we can apply. So, a good way to test the feedback system is by stepping the desired set point to a few speed values and observing how well the output speed tracks the set point. So, using the staircase signaling block, we create the set points signal where every four seconds, the set point changes abruptly to a few different values and we're observing the plots for the desired set point and the measured speed as shown in Figure 1.53. Let's run the file, while observing the plot, you can see that it tracks the set point speed very well and it also meets our design right here in regard to the settling time, overshoot, and steady-state error as shown in Figure 1.54.

So, in simulation, our results look good. Now, let's go implement this exact same test on the Arduino setup. So, here is the Arduino script.

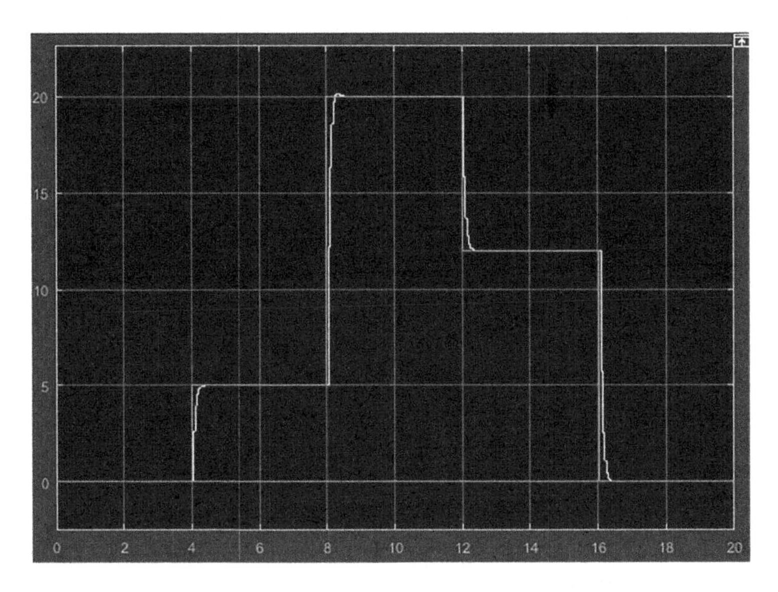

FIGURE 1.54 Plot of tracking the set point speed.

```
int T = 50;                  // sampling time T = 50ms
unsigned long previous_time, current_time;
// variables to hold time values
float encoder_count = 0;     // initialize encoder
count to 0
boolean encoder_A, encoder_B;
byte state, statep;
float previous_angle, current_angle = 0;
// set current and previous angle //values to 0
float output_speed = 0;              // initialize
output speed to 0
float supply_voltage = 12;       // the supply
voltage for our PWM motor driver //is 12V, so we
set this to 12
float elapsed_time = 0;
float setpoint;     // declare setpoint variable
float e, e_1, u, u_1 = 0;        // delcare e[n],
e[n-1], u[n], and u[n-1]
```

```
const int pwm_output = 7;        // label digital
pin 7 as the PWM output
void Encoder_State(){      // this function is
used for encoder pulse counting
 encoder_A = digitalRead(3);        // read pulse
 data from encoder's pin A
 encoder_B = digitalRead(2);        // read pulse
 data from encoder's pin B
 if ((encoder_A == HIGH) && (encoder_B == HIGH))
 state = 1;
 if ((encoder_A == HIGH) && (encoder_B == LOW))
 state = 2;
 if ((encoder_A == LOW) && (encoder_B == LOW))
 state = 3;
 if ((encoder_A == LOW) && (encoder_B == HIGH))
 state = 4;
 switch (state)
 {
  case 1:
  {
   if (statep == 2) encoder_count--;
   if (statep == 4) encoder_count++;
   break;
  }
  case 2:
  {
   if (statep == 1) encoder_count++;
   if (statep == 3) encoder_count--;
   break;
  }
  case 3:
  {
   if (statep == 2) encoder_count++;
   if (statep == 4) encoder_count--;
   break;
  }
  default:
  {
```

```
    if (statep == 1) encoder_count--;
    if (statep == 3) encoder_count++;
    break;
   }
  }
  statep = state;
 }
 void get_speed(){
  current_angle = (encoder_count*360.0)/(1632.67);
  // 1632.67 is the total //number of counts per
  revolution. Because it's a 48 CPR encoder and a
  34.1 //gearmotor, you multiply 48 x 34.1 =
  1632.67
  output_speed = 0.0175*((current_angle - previous_
  angle)/0.05);          // 0.0175 rad/s = 1 degree.
  The formula is essentially calculating theta_dot
  // or dteta/dT, where dT = 0.05, is the sampling
  time in seconds
  previous_angle = current_angle;
 }
 void step_setpoint(){      // a stair-case setpoint
 input signal to test the //transient response of
 the control loop
  setpoint = 0;
  if (current_time >= 4000 && current_time <=
  8000){
    setpoint = 5;       // after 4 seconds,
    setpoint = 5 rad/s for 4 seconds
  }
  if (current_time > 8000 && current_time <=
  12000){
    setpoint = 20;          // setpoint = 20 rad/s
    for 4 seconds
  }
  if (current_time > 12000 && current_time <=
  16000){
    setpoint = 12;          // setpoint = 12 rad/s
    for 4 seconds
  }
```

```
 if (current_time > 16000){
   setpoint = 0;
 }
}
void setup(){
 Serial.begin(9600);
 previous_time = millis();
 pinMode(2,INPUT);           // set digital pin 2
 as an input
 pinMode(3,INPUT);           // set digital pin 3
 as an input
 attachInterrupt(digitalPinToInterrupt(2),
 Encoder_State, CHANGE);
 attachInterrupt(digitalPinToInterrupt(3),
 Encoder_State, CHANGE);
 encoder_A = digitalRead(3);
 encoder_B = digitalRead(2);
 if ((encoder_A == HIGH) && (encoder_B == HIGH))
 statep = 1;
 if ((encoder_A == HIGH) && (encoder_B == LOW))
 statep = 2;
 if ((encoder_A == LOW) && (encoder_B == LOW))
 statep = 3;
 if ((encoder_A == LOW) && (encoder_B == HIGH))
 statep = 4;
 pinMode(pwm_output, OUTPUT);        // set pwm_
 output or pin 7 as the PWM signal's output pin
}
void loop() {
current_time = millis();    // store the elapsed
time in milliseconds
int delta_time = current_time - previous_time;
// calculate the delta //time or time difference
in milliseconds
if (delta_time > T){          // if the time
difference is equal to 50 ms, execute the if
statement. Doing so, keeps a consistent sampling
interval of //50ms
```

```
 elapsed_time = current_time/1000.0;  // stores
 the elapsed time in seconds
 step_setpoint();            // call out desired
 setpoint speed signal
 get_speed();              // obtain our output_
 speed signal in rad/s
 e = setpoint - output_speed;       // calculate
 the error e[n]
 u = u_1 + 0.555*e - 0.299*e_1;     // calculate
 the control signal u[n], this is our difference
 equation derived from C(z)
 if (u >= 12.0){       // if statement is used to
 saturate the control signal value between 0(V)
 and 12(V)
   u = 12;
   }
 else if(u <= 0){
   u = 0;
   }
 analogWrite(pwm_output,(u/supply_voltage)*255.0);
 // maps the control signal values linearly to
 PWM duty cycle %'s. 0 - 12 correlates to 0%
 - 100%.
 u_1 = u;       / store u[n] --> u{n-1] for next
 iteration
 e_1 = e;      // store e[n] --> e{n-1] for next
 iteration
 Serial.print(elapsed_time);      // Serial print
 the elapsed time(s) and the output_speed or
 measured speed(rad/s)
 Serial.print("      ");
 Serial.println(output_speed);
 previous_time = current_time;        // store
 current_time --> previous_time for next loop
 iteration
 }
}
```

You can see that the controller is implemented as a difference equation as we expected inside the void loop function. You can also see that we've declared an input set point function (step_setpoint) that mimics the input set point signal from the simulation. We've also included some code to saturate or limit the control output between 0 and 12. So, let's monitor the set point signal and the measured output speed on the Serial Plotter and observe the DC gearmotor's behavior as depicted in Figure 1.55.

So, it seems to be working as expected. You'll notice that there's a lot of noise in the measured output. This is mainly due to the noise in the encoder speed measurement, but again it's very minimal and the overall transient and steady-state performance is pretty good. With built-in encoders, we don't really expect them

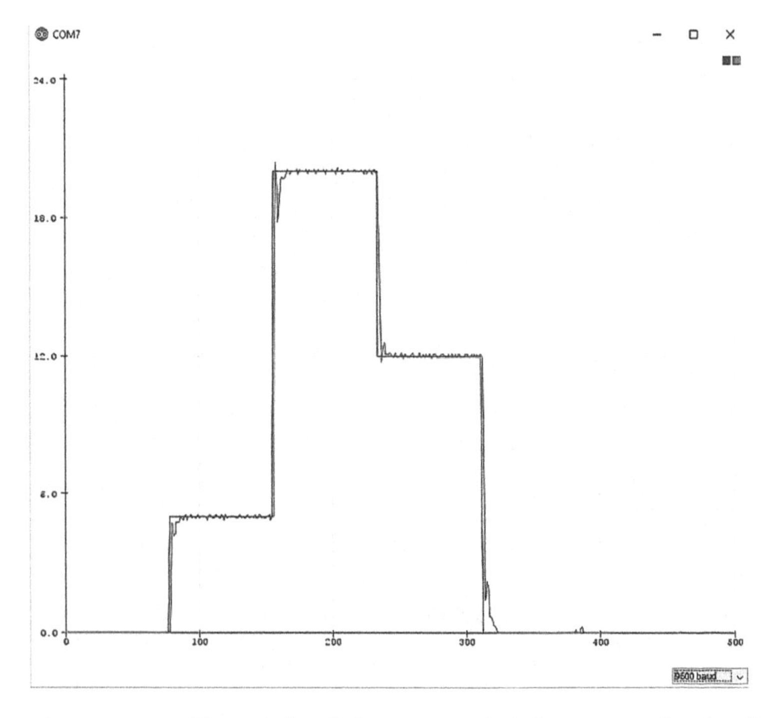

FIGURE 1.55 Observe the DC gearmotor's behavior on the Serial Plotter.

to have very precise or noiseless performance, but overall, this control structure works and does exactly what we expected it to. So, we did the MATLAB Simulink simulation, and then we did the Arduino implementation on the real DC gearmotor. So now, let's compare the two results with one another and see how closely they match or mismatch. We'll modify our Arduino script to output the elapsed time and the output speed and collect 17 seconds with the data using the Serial Monitor.

Then on MATLAB, we'll create a variable called Measured_Speed and enter in two columns of that, this elapsed time (column 1) and measured output speed (column 2) data as shown in Figure 1.56.

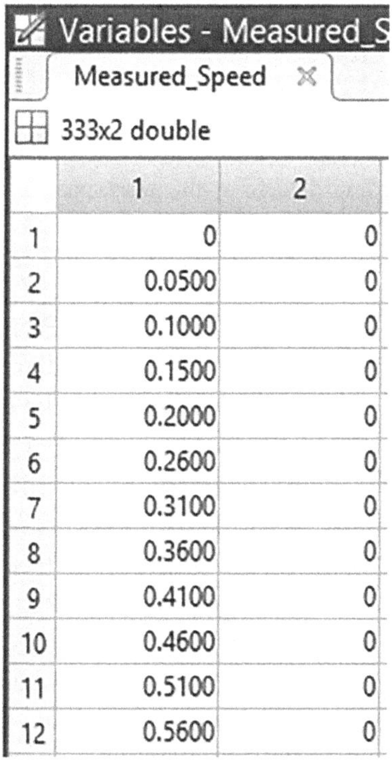

Variables - Measured_S

Measured_Speed

333x2 double

	1	2
1	0	0
2	0.0500	0
3	0.1000	0
4	0.1500	0
5	0.2000	0
6	0.2600	0
7	0.3100	0
8	0.3600	0
9	0.4100	0
10	0.4600	0
11	0.5100	0
12	0.5600	0

FIGURE 1.56 Creating a variable string with this measured output speed data.

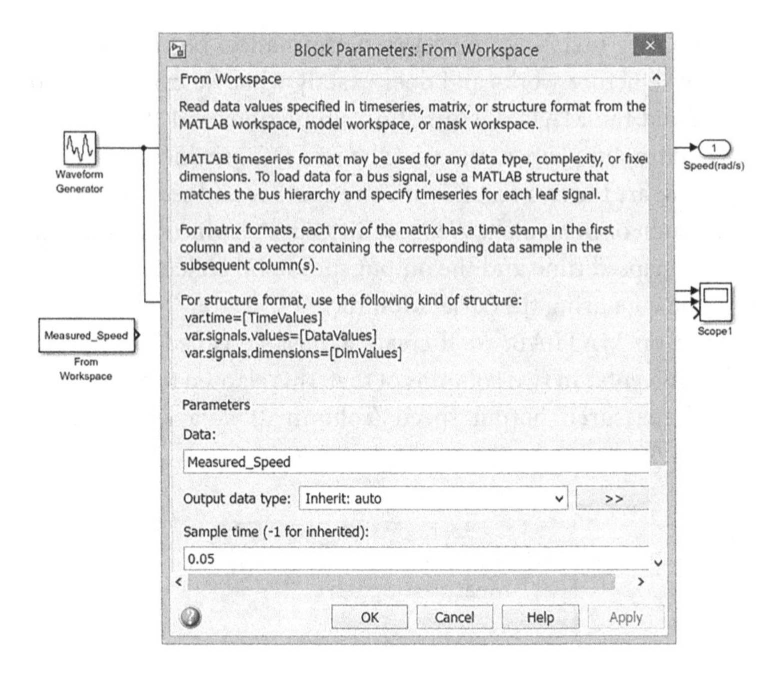

FIGURE 1.57 Feeding data from the workspace and plot against the simulated data – Part 1.

Then on Simulink, we'll feed this data to From the Workspace block and fill in its Data as Measured_Speed and set the Sample time as 0.05 and plot on the scope against the simulated output data and the simulated set point as shown in Figures 1.57 and 1.58, respectively.

As shown in Figure 1.59, the theoretical result and measured result match very well with the exception of the noise on the measure data and you can see that even the transient parts of the measure data match pretty close with the simulated data.

So, this comparison proves three things. First, the transfer function we approximated for our actual DC gearmotor was a good enough approximation to use for our controller design step

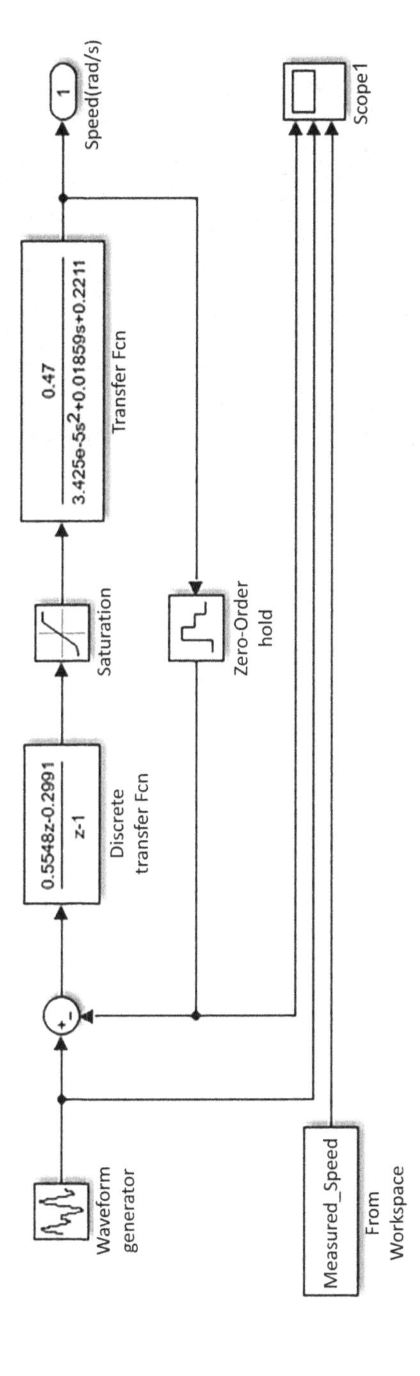

FIGURE 1.58 Feeding data from the workspace and plot against the simulated data – Part 2.

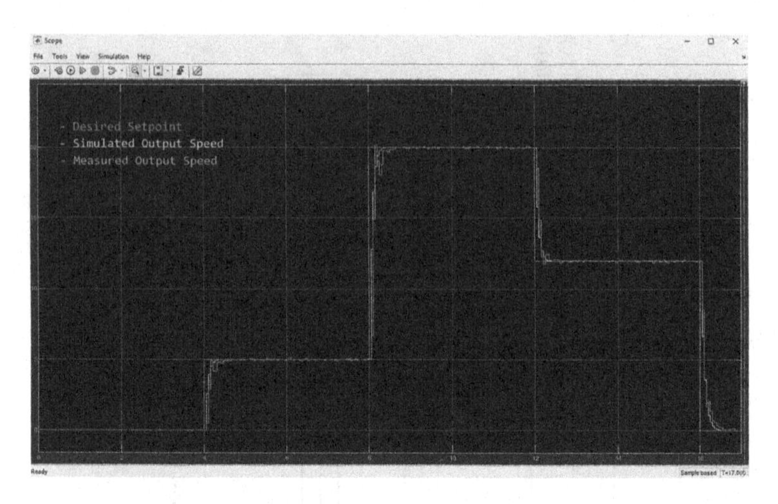

FIGURE 1.59 Comparison of the theoretical result and measured result.

process. Second, the closed-loop control structure on Simulink is a very accurate simulation of the real physical closed-loop system, and, third, the six-step controller design process is theoretically and practically a sound method to use to design a simple digital controller for your continuous-time system, so that is all.

CHAPTER 2

Theoretical Overview of the Buck Converter

2.1 BUCK CONVERTER DESIGN

In this chapter, we will cover the basic design of a buck converter and implement closed-loop control for it. This system has both voltage and current sensors. We do not dive deeply into theoretical analysis of buck converters, because this book is practically oriented. So, we will not consider the derivation of the buck converter equations and we will just present them. The schematic of the basic buck converter is shown in Figure 2.1.

We have an input voltage, V_{IN}, a MOSFET, Q, a diode, D, an inductor, L, a capacitor, C, and a resistive load. The load can be active, for example, a battery or other type of loads, but in this case, a resistor has been used for analysis. We have also the output current, I_o, inductor current, I_L, and output voltage, V_o. The parameters of the buck converter which are constant and designed for a particular case are listed in Table 2.1.

DOI: 10.1201/9781003541356-2

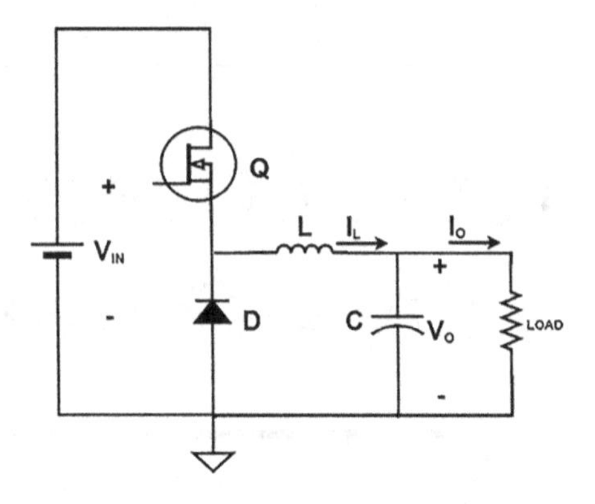

FIGURE 2.1 The schematic of basic buck converter.

TABLE 2.1 Parameters of the Buck Converter

Parameters	Values
V_{in}	$24\ V$
V_{out}	$16\ V$
$I_{\text{out}}(=i_L)$	$2.133\ A$
R_{out}	$7.5\ \Omega$
$\Delta V_{O,\text{pk}-\text{pk}}$	$0.02\ V_{\text{out}}$
$\Delta I_{L,\text{pk}-\text{pk}}$	$0.1\ I_{\text{out}}$
r_{ESR}	$3\ m\Omega$
r_L	$300\ m\Omega$
V_{fwd}	$1.15\ V$
$V_{\text{ds, sat}}$	$0.075\ V$
L	$200\ \mu H$
C	$40\ \mu F$
f_{sw}	$100\ kHz$

For the sake of analysis, we choose the inductor current ripple to be 10% of inductor current, using the following equations for duty cycle, D, and inductor, L, we get the duty cycle and inductance values:

$$D = \frac{V_{\text{fwd}} + i_L r_L + V_{\text{out}}}{V_{\text{in}} + V_{\text{fwd}} - V_{\text{ce,sat}}} = 0.7094 \tag{2.1}$$

$$L \geq \frac{\left(V_{\text{in}} - V_{\text{ce,sat}} - i_L r_L - V_{\text{out}}\right) \times D}{\Delta I_{L,\text{pk-pk}} f_{\text{sw}}} = 243.3\,\mu H \tag{2.2}$$

Because of the size of the inductor, we choose the inductor L to be 200 μH and let the inductor current ripple to be 15% of inductor current. Additionally, the output voltage ripple, which is 2% of the output voltage, and the actual capacitor value (40 μF) significantly exceed the calculated value for capacitor C (0.83 μF), as shown in the following equation:

$$C \geq \frac{\Delta I_{L,\text{pk-pk}}}{8 \Delta V_{o,\text{pk-pk}} f_{\text{sw}}} = 0.83\,\mu F \tag{2.3}$$

Then we have the following small signal transfer function which is crucial in designing the buck converter [5]:

$$\frac{\hat{i}_o(s)}{\hat{d}(s)} = \frac{\left(1 + sCr_{\text{ESR}}\right)\left(V_{\text{in}} + V_{\text{fwd}} - V_{\text{ds,sat}}\right)}{\left(LCR + LCr_{\text{ESR}}\right)s^2 + \left(CRr_L + Cr_{\text{ESR}} r_L + CRr_{\text{ESR}} + L\right)s + r_L + R_{\text{out}}} \tag{2.4}$$

We should note that these equations cannot be used to other converters. Also, the output voltage to duty cycle small signal transfer function is as below:

$$\frac{\hat{v}_o(s)}{\hat{d}(s)} = \frac{R_{\text{out}}\left(1 + sCr_{\text{ESR}}\right)\left(V_{\text{in}} + V_{\text{fwd}} - V_{\text{ce,sat}}\right)}{\left(LCR + LCr_{\text{ESR}}\right)s^2 + \left(CRr_L + Cr_{\text{ESR}} r_L + CRr_{\text{ESR}} + L\right)s + r_L + R_{\text{out}}} \tag{2.5}$$

Notice that the parasitic parameters, that is, equivalent series resistance of capacitor and inductor resistance, are denoted by r_{ESR}, and r_L respectively. If we substitute the parameters in Table 2.1 in the above equations, we will have the following relations:

$$\frac{\hat{i}_o(s)}{\hat{d}(s)} = \frac{3.009\times10^{-6}s+25.07}{6.003\times10^{-8}s^2+0.0002912s+7.801} \tag{2.6}$$

$$\frac{\hat{v}_o(s)}{\hat{d}(s)} = \frac{2.257\times10^{-5}s+188.1}{6.003\times10^{-8}s^2+0.000291s+7.801} \tag{2.7}$$

For designing a PI controller for current control loop or voltage control loop, we can use MATLAB. In the previous chapter, we used MATLAB functions (pidTuner and c2d tools) for designing a digital PID controller for our control loop. To design a PI controller for the current control loop using MATLAB, the following code can be applied:

```
s = tf('s');
Gid = (3e-6*s+25.07)/(6e-8*s^2+0.000291*s+7.801);
pidTuner(Gid);
```

By executing the above code, the PID tuner window is popped up. We select the type of controller as PI and its form as Standard and then regulate the Response Time and Transient Behavior sliders to get desired Step Plot. Then, if we click on Show Parameters button, we can see our controller parameters. Finally, we click on the Export button and give it a name for our PI controller. In this case, we chose 'Cont' as its name and click on the OK button as shown in Figure 2.2.

Now, in the command window, if we type the name ('Cont') and then press the Enter key, you can get the controller transfer function as the following equation:

$$G_c(s) = K_p\left(1+\frac{1}{T_i s}\right) = K_p + \frac{\left(\frac{K_p}{T_i}\right)}{s} = 3.11\times10^{-5} + \frac{120.5}{s} \tag{2.8}$$

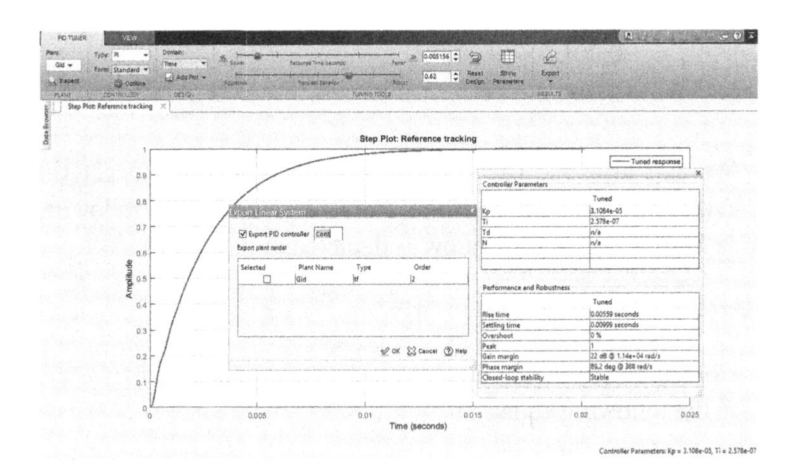

FIGURE 2.2 The PID tuner window.

$$K_p = 3.11\times10^{-5}, T_i = 2.58\times10^{-7} \tag{2.9}$$

If you want to implement analog controller using op-amp, the above equation is enough. However, for use in a digital controller, the following substitutions must be applied (Tustin's method):

$$s = \frac{2}{T_s}\frac{(z-1)}{(z+1)} = \frac{2}{T_s}\frac{(1-z^{-1})}{(1+z^{-1})} \tag{2.10}$$

where T_S is the sampling time and is chosen as 100 μs, so we have:

$$\begin{aligned} G_c(z) &= K_p\left(1+\frac{T_s(z+1)}{2T_i(z-1)}\right) \\ &= \frac{K_p(z-1)+K_pT_s(z+1)/(2T_i)}{(z-1)} \\ &= \frac{\left(K_p+K_pT_s/(2T_i)\right)z+K_pT_s/(2T_i)-K_p}{z-1} \\ &= \frac{0.0061z+0.006}{z-1} \end{aligned} \tag{2.11}$$

$$G_c(z) = \frac{0.006z^{-1} + 0.0061}{-z^{-1} + 1} \tag{2.12}$$

We can also use the c2d function in MATLAB to get the discrete-time transfer function from $G_c(s)$ using the following code in the command window as depicted in Figure 2.3.

```
c2d(Cont,100e-6,'tustin')
```

Correlating the equation above with the standard form, we will have the following equation:

$$G_c(z) = \frac{B_3z^{-3} + B_2z^{-2} + B_1z^{-1} + B_0}{-A_3z^{-3} - A_2z^{-2} - A_1z^{-1} + 1} = \frac{y_n}{e_n} \tag{2.13}$$

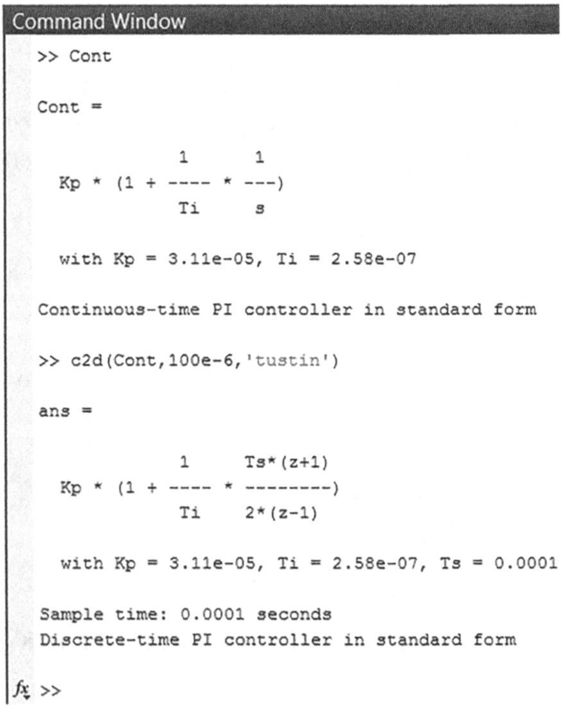

```
Command Window
>> Cont

Cont =

               1      1
  Kp * (1 + ---- * ---)
              Ti      s

  with Kp = 3.11e-05, Ti = 2.58e-07

Continuous-time PI controller in standard form

>> c2d(Cont,100e-6,'tustin')

ans =

               1     Ts*(z+1)
  Kp * (1 + ---- * --------)
              Ti     2*(z-1)

  with Kp = 3.11e-05, Ti = 2.58e-07, Ts = 0.0001

Sample time: 0.0001 seconds
Discrete-time PI controller in standard form

fx >>
```

FIGURE 2.3 The PI controller transfer function in continuous- and discrete-time domain.

$$A_1 = 1,\ B_0 = 0.0061,\ B_1 = 0.006 \tag{2.14}$$

The equivalent equation is given as follows:

$$y_n = A_3 y_{n-3} + A_2 y_{n-2} + A_1 y_{n-1} + B_3 e_{n-3} + B_2 e_{n-2} + B_1 e_{n-1} + B_0 e_n \tag{2.15}$$

To design a PI controller for voltage control loop, the same as the current control loop, we can use pidTuner and c2d tools in MATLAB. To design a PI controller for the voltage control loop using MATLAB, the following code can be applied.

```
s = tf('s');
Gvd = (2.257e-5*s+188.1)/(6e-8*s^2+0.000291*s
+7.801);
pidTuner(Gvd);
```

By executing the above code, the PID tuner window is popped up. We select the type of controller as PI and its form as Standard and then regulate the Response Time and Transient Behavior sliders to get desired Step Plot. Then, if we click on Show Parameters button, we can see our controller parameters. Finally, we click on the Export button and give it a name for our PI controller, in this case, we chose 'Cv' as its name and click on the OK button as shown in Figure 2.4.

Now, in the command window, if we type the name ('Cv') and then press the Enter key, you can get the controller transfer function as the following equation:

$$G_c(s) = K_p\left(1 + \frac{1}{T_i s}\right) = K_p + \frac{\left(\frac{K_p}{T_i}\right)}{s} = 4.14\times10^{-6} + \frac{16.05}{s} \tag{2.16}$$

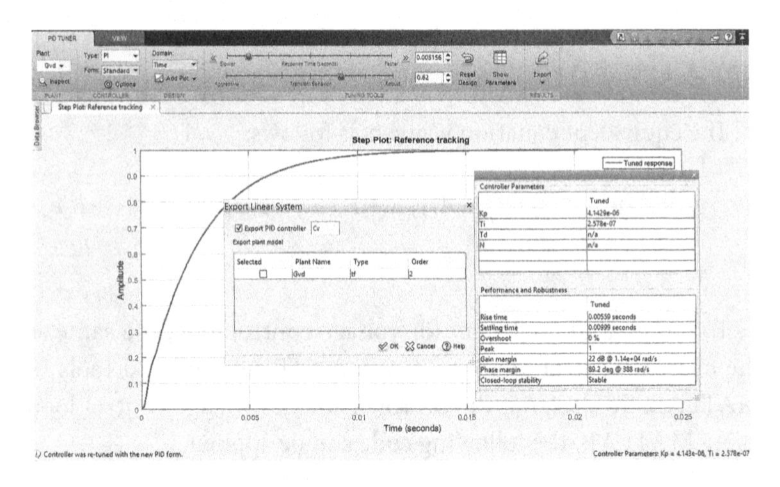

FIGURE 2.4 The PID tuner window.

$$K_p = 4.14 \times 10^{-6}, T_i = 2.58 \times 10^{-7} \tag{2.17}$$

If you want to implement analog controller using op-amp, the above equation is enough. However, for use in a digital controller, the following substitutions must be applied (Tustin's method):

$$s = \frac{2}{T_s}\frac{(z-1)}{(z+1)} = \frac{2}{T_s}\frac{(1-z^{-1})}{(1+z^{-1})} \tag{2.18}$$

where T_S is the sampling time and is chosen as 100 μs, so we have:

$$\begin{aligned} G_c(z) &= K_p\left(1+\frac{T_s(z+1)}{2T_i(z-1)}\right) \\ &= \frac{K_p(z-1)+K_pT_s(z+1)/(2T_i)}{(z-1)} \\ &= \frac{\left(K_p+K_pT_s/(2T_i)\right)z+K_pT_s/(2T_i)-K_p}{z-1} \\ &= \frac{0.00081z+0.0008}{z-1} \end{aligned} \tag{2.19}$$

```
Command Window
>> Cv

Cv =

             1     1
  Kp * (1 + ---- * ---)
             Ti    s

  with Kp = 4.14e-06, Ti = 2.58e-07

Continuous-time PI controller in standard form

>> c2d(Cv,100e-6,'tustin')

ans =

             1      Ts*(z+1)
  Kp * (1 + ---- * --------)
             Ti     2*(z-1)

  with Kp = 4.14e-06, Ti = 2.58e-07, Ts = 0.0001

Sample time: 0.0001 seconds
Discrete-time PI controller in standard form

fx >>
```

FIGURE 2.5 The PI controller transfer function in continuous- and discrete-time domain.

$$G_c\left(z\right)=\frac{0.0008z^{-1}+0.00081}{-z^{-1}+1} \tag{2.20}$$

We can also use the c2d function in MATLAB to get the discrete-time transfer function from $G_c(s)$ in the command window, with the code below as depicted in Figure 2.5.

```
c2d(Cv,100e-6,'tustin')
```

Correlating the equation above with the standard form, we will have the following equation:

$$G_c\left(z\right)=\frac{B_3z^{-3}+B_2z^{-2}+B_1z^{-1}+B_0}{-A_3z^{-3}-A_2z^{-2}-A_1z^{-1}+1}=\frac{y_n}{e_n} \tag{2.21}$$

$$A_1 = 1, B_0 = 0.00081, B_1 = 0.0008 \tag{2.22}$$

The equivalent equation is given as follows:

$$y_n = A_3 y_{n-3} + A_2 y_{n-2} + A_1 y_{n-1} + B_3 e_{n-3} + B_2 e_{n-2} + B_1 e_{n-1} + B_0 e_n \tag{2.23}$$

In continuous conduction mode (CCM), we keep the inductor current ripple very low (usually lower than 20%). We select the inductor, L, lower than the calculated value in Equation (2.2), for practical reasons. Now, we are going to introduce another method for controller design. Here, we are going to set the controller as simple as possible. We use the phase margin to design the controller. So, we set up the phase margin that we wanted and designed the controller to meet that phase margin at a specific crossover frequency. The crossover frequency in control systems refers to the frequency at which the magnitude of the open-loop transfer function is equal to 1 or 0 dB. The switching frequency, on the other hand, refers to the frequency at which a switching device (such as a transistor in a power converter) turns on and off. There is no direct relationship between the crossover frequency and the switching frequency in control systems. These are two separate concepts that relate to different aspects of a system. However, the choice of switching frequency in a power converter can indirectly affect the crossover frequency of the control system that is regulating the converter. For example, a higher switching frequency in a power converter may allow for faster response times in the control system, which could potentially impact the crossover frequency. Ultimately, the crossover frequency is determined by the design of the control system itself and is not directly tied to the switching frequency. In this case, the crossover frequency is chosen to be $f_c = 2$ kHz. So, for the current controller, we have the following equations:

$$s = j\omega_c = j2\pi f_c = j12566.4 \tag{2.24}$$

$$(2.6) \rightarrow \frac{\hat{i}_o(j12566.4)}{\hat{d}(j12566.4)} = 6.236\angle -114.6^{\circ} \tag{2.25}$$

$$\text{Gain} = 6.236 \tag{2.26}$$

$$\text{Phase} = -114.6^{\circ} \tag{2.27}$$

The following equations are used to determine the compensation required to do that we specify a phase margin in crossover frequency and calculate the required compensation phase:

$$\varphi_{\text{PM}} = 59^{\circ} \tag{2.28}$$

$$\varphi_{\text{crossover}} + \varphi_{\text{comp}} + 180^{\circ} = \varphi_{\text{PM}} \tag{2.29}$$

$$\varphi_{\text{comp}} = \varphi_{\text{PM}} - \varphi_{\text{crossover}} - 180^{\circ} \tag{2.30}$$

$$\varphi_{\text{comp}} = 59^{\circ} - \left(-114.6^{\circ}\right) - 180^{\circ} = -6.4^{\circ} \tag{2.31}$$

We usually choose phase margin between 40 and 60 degrees. However, we do not want the system to be unstable, so we have selected higher phase margin in this case which is 59 degrees. If the compensator phase is positive, you will need a lead-lag compensator. If the compensator phase is negative, you will need a lag compensator, which in this case is a PI controller. Using the above equations, a lag compensator is adequate, so the PI compensator is of the form:

$$G_c(s) = K_p\left(\frac{\omega_i}{s} + 1\right) \tag{2.32}$$

$$G_c(j\omega_c) = K_p\left(\frac{\omega_i}{j\omega_c} + 1\right) \tag{2.33}$$

$$\angle G_c(j\omega_c) = -\tan^{-1}\left(\frac{\omega_i}{\omega_c}\right) \tag{2.34}$$

$$-6.4^\circ = -\tan^{-1}\left(\frac{\omega_i}{\omega_c}\right) \tag{2.35}$$

$$\omega_i = \omega_c \times \tan\left(6.4^\circ\right) \tag{2.36}$$

$$\omega_i = 12566.4 \times \tan\left(6.4^\circ\right) = 1409.55\,\text{rad/s} \tag{2.37}$$

The value for K_p is found using the following equations:

$$\left|G_c(j\omega_c)\right| = K_p\sqrt{\left(\frac{\omega_i}{\omega_c}\right)^2 + 1} = \frac{1}{\text{Gain}} = \frac{1}{6.236} \tag{2.38}$$

$$K_p = \frac{1/6.236}{\sqrt{\left(\tan\left(6.4^\circ\right)\right)^2 + 1}} = 0.159 \tag{2.39}$$

Since, in the crossover frequency, the gain of compensated system should be equal to 0 dB or 1. Therefore, the controller transfer function is of the form:

$$G_c(s) = K_p\left(\frac{\omega_i}{s} + 1\right) = 0.159\left(\frac{1409.55}{s} + 1\right) = 0.159 + \frac{224.12}{s} \tag{2.40}$$

We will have the identical crossover frequency for voltage control loop as well. The following equations are applied for the voltage controller. At 2 kHz, the following info is obtained:

$$s = j\omega_c = j2\pi f_c = j12566.4 \tag{2.41}$$

$$(2.7) \rightarrow \frac{\breve{v}_o(j12566.4)}{\breve{d}(j12566.4)} = 46.79\angle -114.6^\circ \tag{2.42}$$

$$\text{Gain} = 46.79 \tag{2.43}$$

$$\text{Phase} = -114.6^\circ \tag{2.44}$$

The following equations are used to determine the compensation required to do that we specify a phase margin in crossover frequency and calculate the required compensation phase:

$$\varphi_{\text{PM}} = 59^\circ \tag{2.45}$$

$$\varphi_{\text{crossover}} + \varphi_{\text{comp}} + 180^\circ = \varphi_{\text{PM}} \tag{2.46}$$

$$\varphi_{\text{comp}} = \varphi_{\text{PM}} - \varphi_{\text{crossover}} - 180^\circ \tag{2.47}$$

$$\varphi_{\text{comp}} = 59^\circ - (-114.6^\circ) - 180^\circ = -6.4^\circ \tag{2.48}$$

We usually choose phase margin between 40 and 60 degrees. However, we do not want the system to be unstable, so we have selected higher phase margin in this case which is 59 degrees. If the compensator phase is positive, you will need a lead-lag compensator. If the compensator phase is negative, you will need a lag compensator, which in this case is a PI controller. Using the above equations, a lag compensator is adequate, so the PI compensator is of the form:

$$G_c(s) = K_p\left(\frac{\omega_i}{s} + 1\right) \tag{2.49}$$

$$G_c(j\omega_c) = K_p\left(\frac{\omega_i}{j\omega_c} + 1\right) \tag{2.50}$$

$$\angle G_c(j\omega_c) = -\tan^{-1}\left(\frac{\omega_i}{\omega_c}\right) \tag{2.51}$$

$$-6.4^\circ = -\tan^{-1}\left(\frac{\omega_i}{\omega_c}\right) \tag{2.52}$$

$$\omega_i = \omega_c \times \tan\left(6.4^o\right) \tag{2.53}$$

$$\omega_i = 12566.4 \times \tan\left(6.4^\circ\right) = 1409.55\,\text{rad/s} \tag{2.54}$$

The value for K_p is found using the following equations:

$$\left|G_c(j\omega_c)\right| = K_p\sqrt{\left(\frac{\omega_i}{\omega_c}\right)^2 + 1} = \frac{1}{\text{Gain}} = \frac{1}{46.79} \tag{2.55}$$

$$K_p = \frac{1/46.79}{\sqrt{\left(\tan\left(6.4^\circ\right)\right)^2 + 1}} = 0.021 \tag{2.56}$$

Since, in the crossover frequency, the gain of compensated system should be 0 dB or 1. Therefore, the controller transfer function is of the form:

$$G_c(s) = K_p\left(\frac{\omega_i}{s} + 1\right) = 0.021\left(\frac{1409.55}{s} + 1\right) = 0.021 + \frac{29.6}{s} \tag{2.57}$$

We can use Google Colab for Bode plots of closed-loop current and voltage control with PI controllers designed by MATLAB pidTuner tool and phase margin-based frequency response calculations. Google Colab is a free cloud-based Jupyter notebook environment that allows users to write and execute Python code in a browser. It provides access to free GPU and TPU resources, making it a popular choice for data scientists and machine-learning researchers to work on computation-intensive tasks. Google Colab also allows for easy sharing and collaboration on projects, the Google Colab code is as below.

```
!pip install control

%matplotlib inline
import numpy as np
import matplotlib.pyplot as plt
from control.matlab import *
Gid = tf([3e-6, 25.07],[7.33e-8, 0.0002912,
7.801])
Gvd = tf([2.257e-5, 188.1],[7.33e-8, 0.000291,
7.801])
print("Gid = ", Gid)
print("Gvd = ", Gvd)
Gid =  <TransferFunction>: sys[0]
Inputs (1): ['u[0]']
Outputs (1): ['y[0]']
```

```
    3e-06 s + 25.07
----------------------------------
7.33e-08 s^2 + 0.0002912 s + 7.801

Gvd =  <TransferFunction>: sys[1]
Inputs (1): ['u[0]']
Outputs (1): ['y[0]']

    2.257e-05 s + 188.1
---------------------------------
7.33e-08 s^2 + 0.000291 s + 7.801
```

The above codes are used for printing the transfer functions of Gid and Gvd as given in Equations (2.6) and (2.7). The code below is used for the Bode plot of Gid as illustrated in Figure 2.6.

```
w = np.logspace(1,5)
mag,phase,omega = bode(Gid,w)
plt.tight_layout()
```

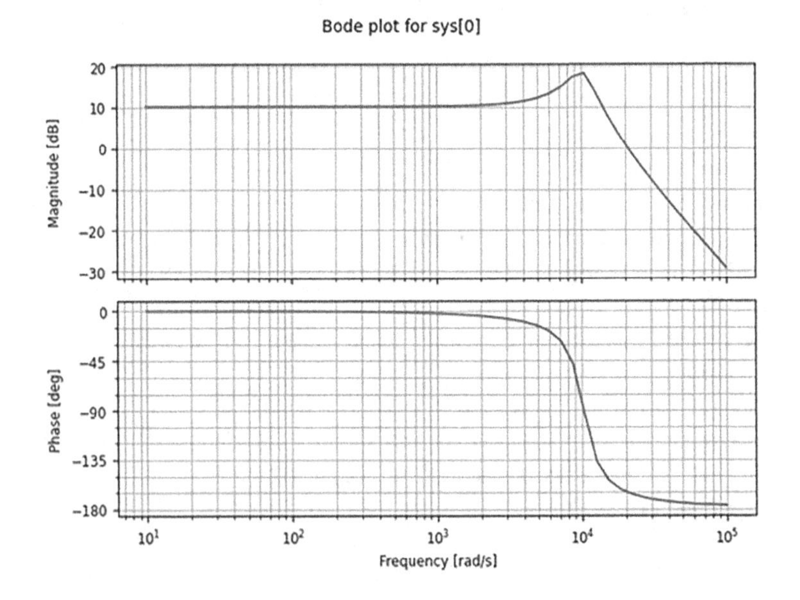

FIGURE 2.6 The Bode plot of Gid.

The Gci_mat is the transfer function of current controller obtained by the pidTuner tool of MATLAB as given in the code below.

```
Gci_mat = tf([3.11e-5, 120.5],[1, 0])
print("Gci_mat = ", Gci_mat)
Gci_mat =  <TransferFunction>: sys[3]
Inputs (1): ['u[0]']
Outputs (1): ['y[0]']

3.11e-05 s + 120.5
------------------
    s
```

The feedback loop transfer function is the product of the plant and controller transfer functions and the following code is used for the Bode plot of it as shown in Figure 2.7.

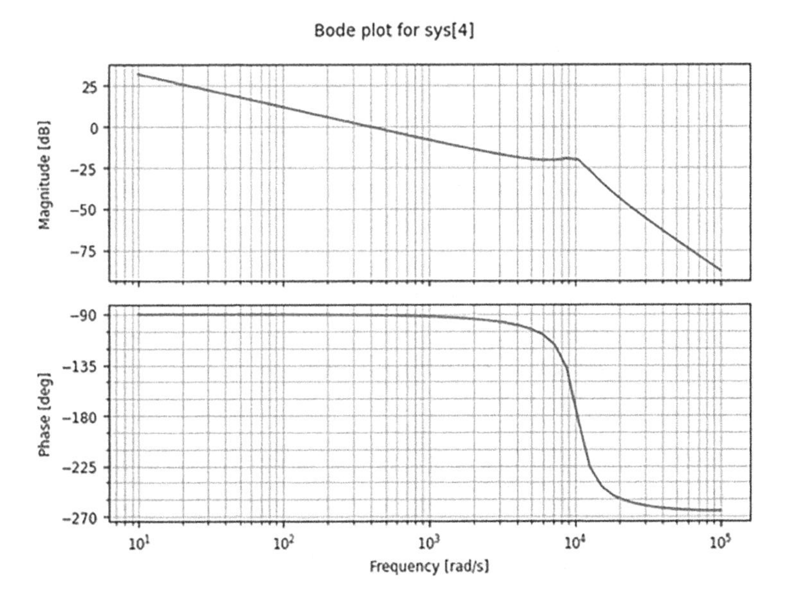

FIGURE 2.7 The Bode plot of Gpci_mat.

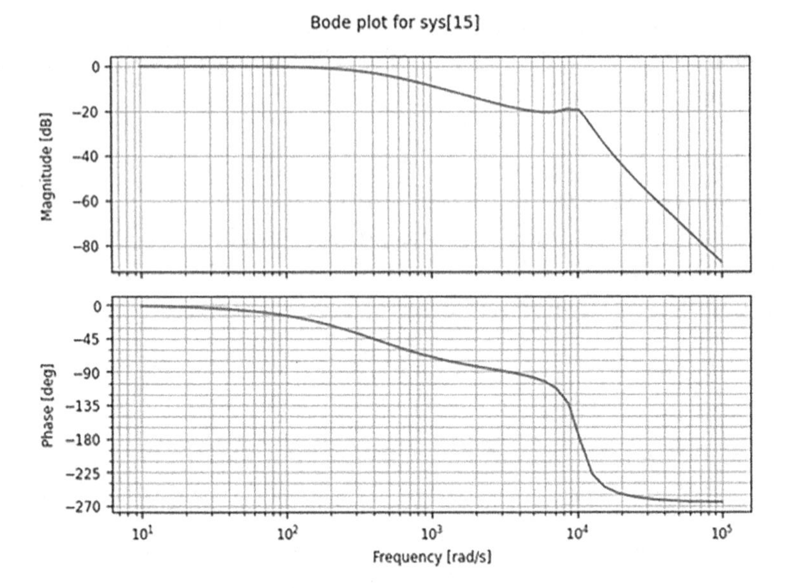

FIGURE 2.8 The Bode plot of Tpci_mat.

```
Gpci_mat = Gid*Gci_mat
w = np.logspace(1,5)
mag,phase,omega = bode(Gpci_mat,w)
plt.tight_layout()
```

The closed-loop transfer function and the code for the Bode plot of it are depicted in Figure 2.8.

```
Tpci_mat = Gpci_mat/(1+Gpci_mat)
w = np.logspace(1,5)
mag,phase,omega = bode(Tpci_mat,w)
plt.tight_layout()
```

The Gci_pm is the transfer function of current controller obtained by the phase margin calculations as given in the code below.

```
Gci_pm = tf([0.159, 224.12],[1, 0])
print("Gci_pm = ", Gci_pm)
```

```
Gci_pm =  <TransferFunction>: sys[10]
Inputs (1): ['u[0]']
Outputs (1): ['y[0]']

0.159 s + 224.1
---------------
      s
```

The feedback loop transfer function is the product of the plant and controller transfer functions and the following code is used for the Bode plot of it as shown in Figure 2.9.

```
Gpci_pm = Gid*Gci_pm
w = np.logspace(1,5)
mag,phase,omega = bode(Gpci_pm,w)
plt.tight_layout()
```

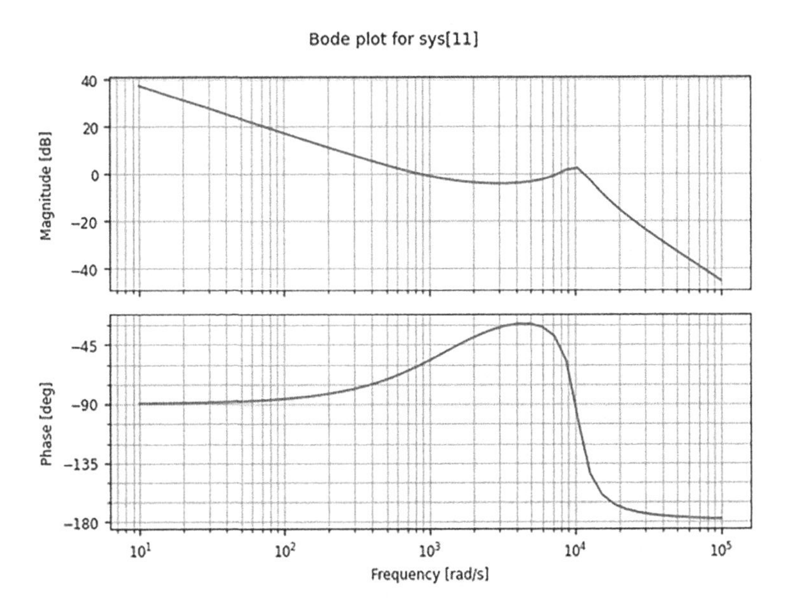

FIGURE 2.9 The Bode plot of Gpci_pm.

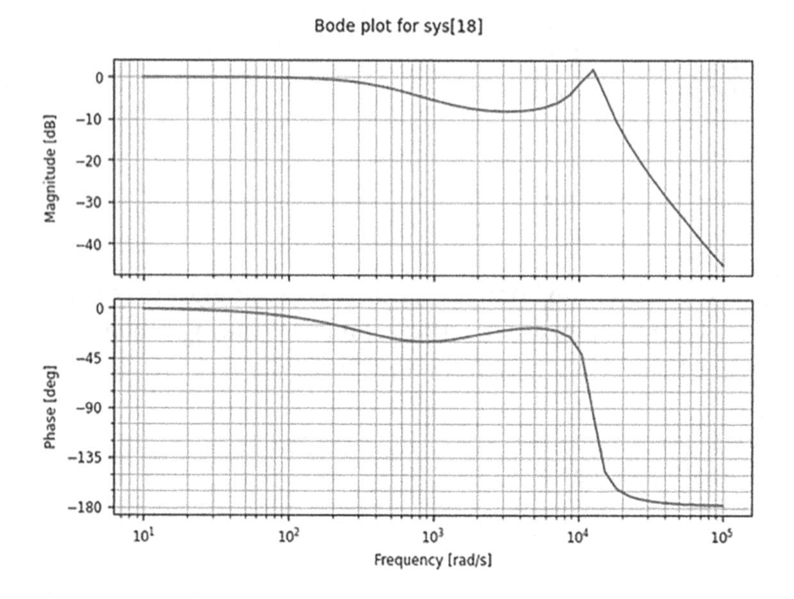

FIGURE 2.10 The Bode plot of Tpci_pm.

The closed-loop transfer function and the code for the Bode plot of it are depicted in Figure 2.10.

```
Tpci_pm = Gpci_pm/(1+Gpci_pm)
w = np.logspace(1,5)
mag,phase,omega = bode(Tpci_pm,w)
plt.tight_layout()
```

The code below is used for the Bode plot of Gvd as illustrated in Figure 2.11.

```
w = np.logspace(1,5)
mag,phase,omega = bode(Gvd,w)
plt.tight_layout()
```

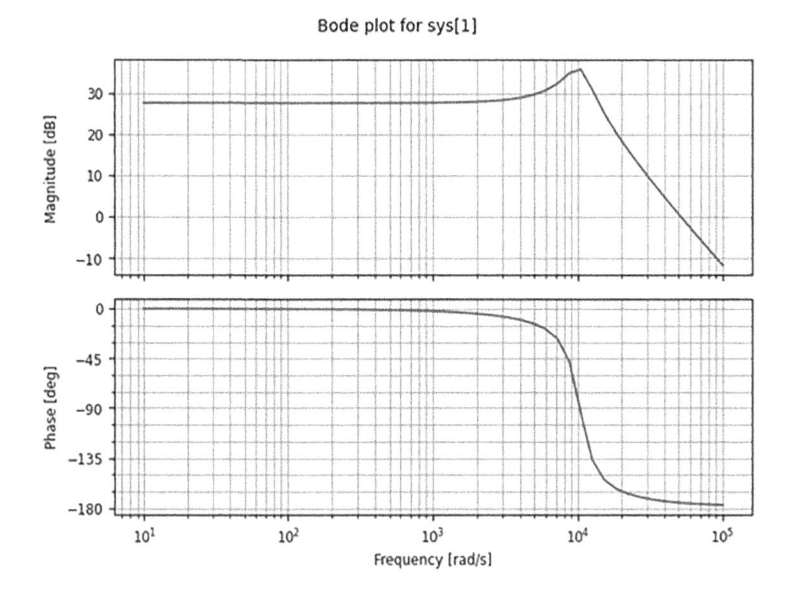

FIGURE 2.11 The Bode plot of Gvd.

The Gcv_mat is the transfer function of current controller obtained by the pidTuner tool of MATLAB as given in the code below.

```
Gcv_mat = tf([4.14e-6, 16.05],[1, 0])
print("Gcv_mat = ", Gcv_mat)
Gcv_mat =  <TransferFunction>: sys[42]
Inputs (1): ['u[0]']
Outputs (1): ['y[0]']

4.14e-06 s + 16.05
------------------
      s
```

The feedback loop transfer function is the product of the plant and controller transfer functions and the following code is used for the Bode plot of it as shown in Figure 2.12.

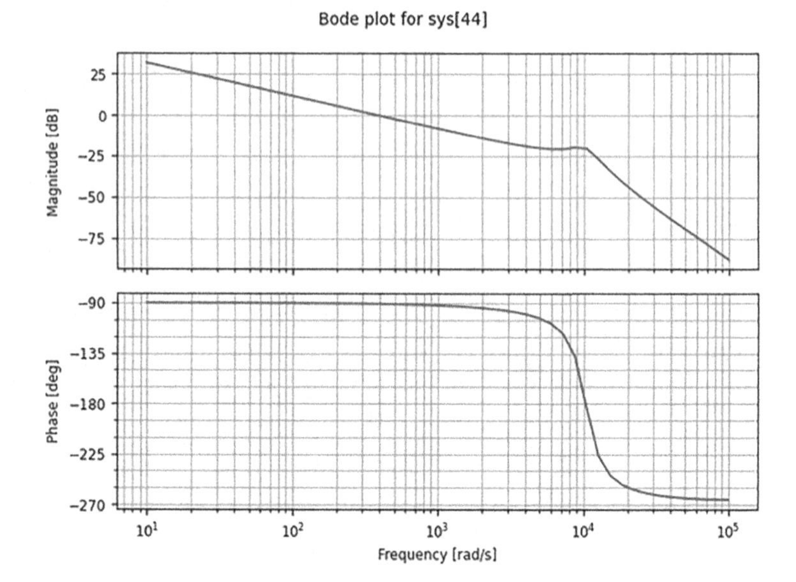

FIGURE 2.12 The Bode plot of Gpcv_mat.

```
Gpcv_mat = Gvd*Gcv_mat
w = np.logspace(1,5)
mag,phase,omega = bode(Gpcv_mat,w)
plt.tight_layout()
```

The closed-loop transfer function and the code for the Bode plot of it are depicted in Figure 2.13.

```
Tpcv_mat = Gpcv_mat/(1+Gpcv_mat)
w = np.logspace(1,5)
mag,phase,omega = bode(Tpcv_mat,w)
plt.tight_layout()
```

The Gcv_pm is the transfer function of current controller obtained by the phase margin calculations as given in the code below.

```
Gcv_pm = tf([0.021, 29.6],[1, 0])
print("Gcv_pm = ", Gcv_pm)
```

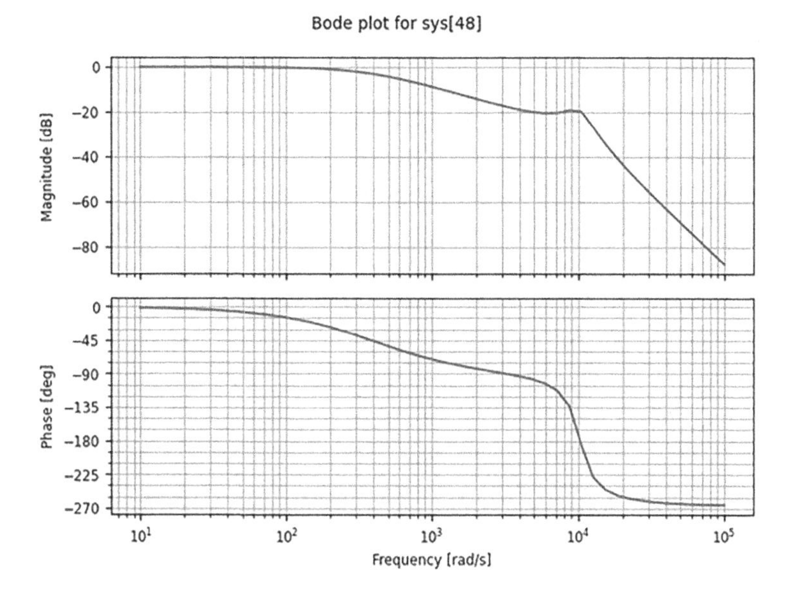

FIGURE 2.13 The Bode plot of Tpcv_mat.

```
Gcv_pm =  <TransferFunction>: sys[50]
Inputs (1): ['u[0]']
Outputs (1): ['y[0]']

0.021 s + 29.6
--------------
     s
```

The feedback loop transfer function is the product of the plant and controller transfer functions and the following code is used for the Bode plot of it as shown in Figure 2.14.

```
Gpcv_pm = Gvd*Gcv_pm
w = np.logspace(1,5)
mag,phase,omega = bode(Gpcv_pm,w)
plt.tight_layout()
```

The closed-loop transfer function and the code for the Bode plot of it are depicted in Figure 2.15.

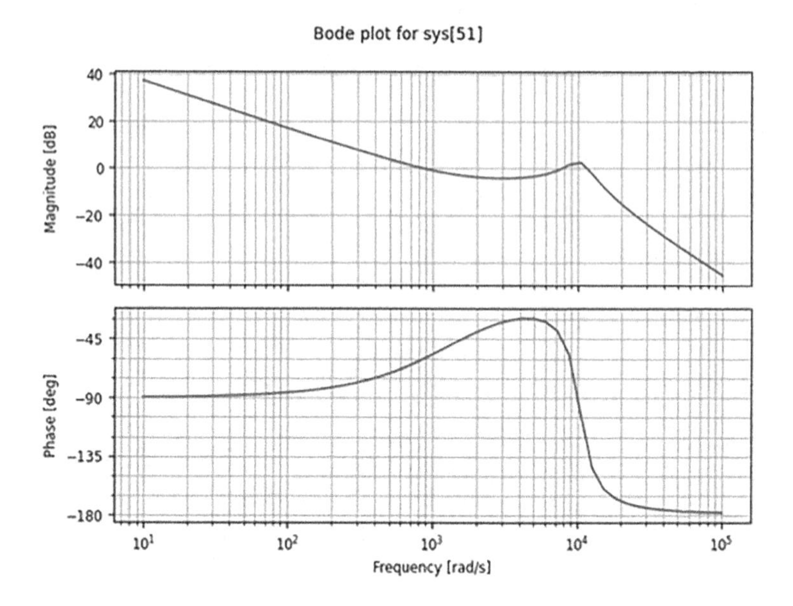

FIGURE 2.14 The Bode plot of Gpcv_pm.

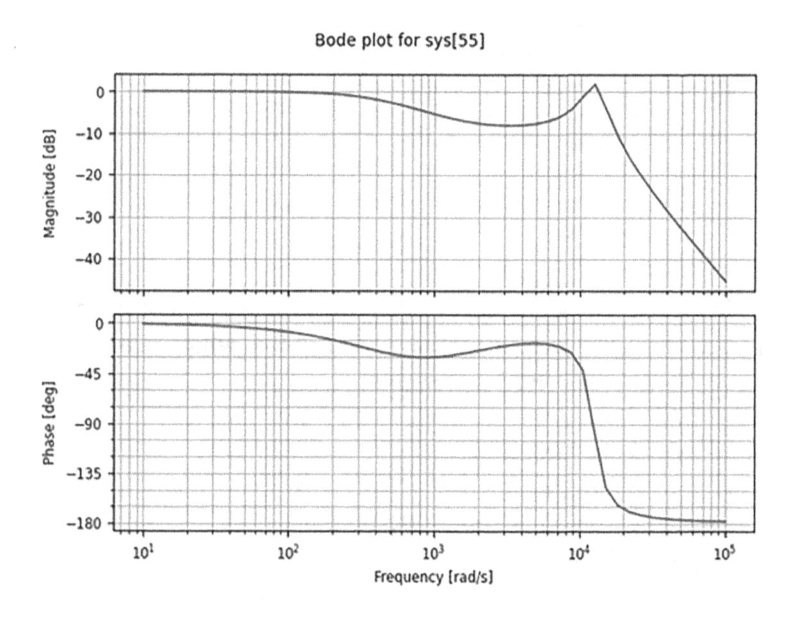

FIGURE 2.15 The Bode plot of Tpcv_pm.

```
Tpcv_pm = Gpcv_pm/(1+Gpcv_pm)
w = np.logspace(1,5)
mag,phase,omega = bode(Tpcv_pm,w)
plt.tight_layout()
```

As you can see in Bode plots in Google Colab for closed-loop current and voltage control using MATLAB pidTuner tool, and phase margin calculations, the designed PI controller for current and voltage control loops is stable. For further analysis, we can compare the operation of designed PI controllers in MATLAB Simulink. As illustrated in Figure 2.16, we are going to compare the step or transient response of closed-loop current and voltage control with PI controllers designed by MATLAB pidTuner tool and phase margin calculations.

The results of transient response of current and voltage control loops with PI controllers designed by MATLAB pidTuner tool and phase margin calculations are plotted on the scopes as shown in Figures 2.17 and 2.18, respectively.

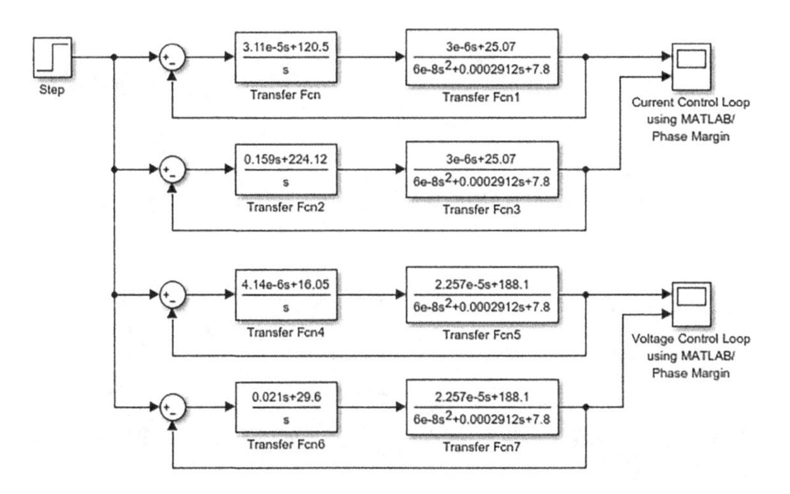

FIGURE 2.16 Comparing the transient response of current and voltage control loops with PI controllers designed by MATLAB pidTuner tool and phase margin calculations.

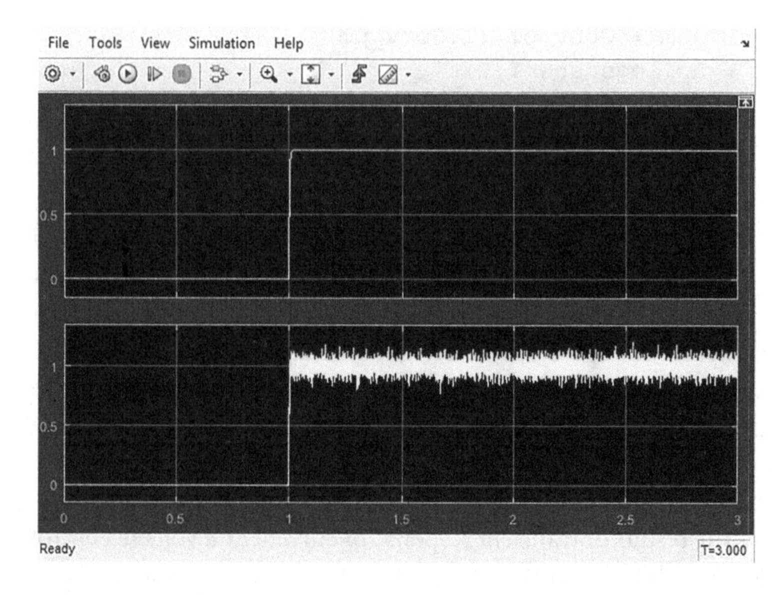

FIGURE 2.17 The step response of current control loop (up: MATLAB, down: Phase Margin).

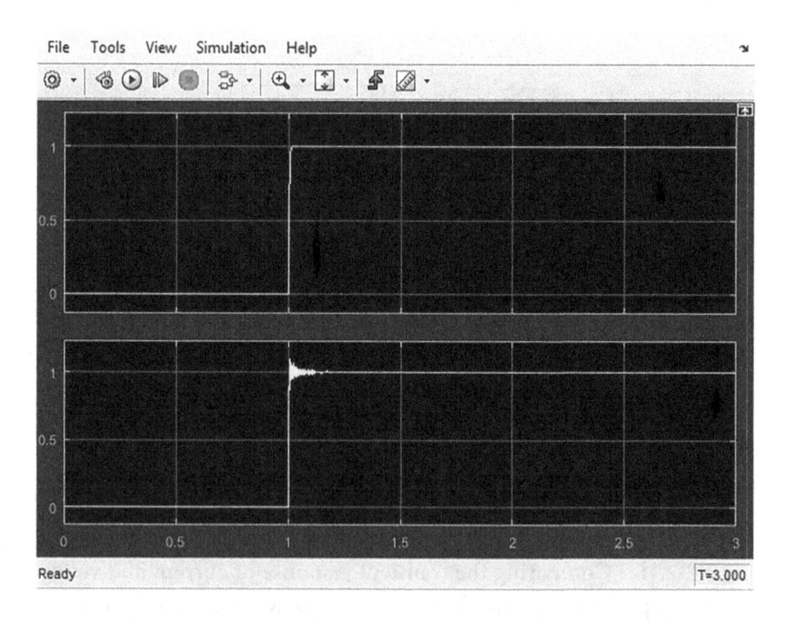

FIGURE 2.18 The step response of voltage control loop (up: MATLAB, down: Phase Margin).

Although the PI controllers with both methods are stable, however, the PI controllers designed by pidTuner tool of MATLAB have better transient response than phase margin calculations. Therefore, we will use the PI controllers designed by pidTuner tool of MATLAB.

2.2 THE CONTROLLER DESIGN WITH OCTAVE

In this section, we are going to discuss about how to obtain the digital controller coefficients for designing closed-loop buck converter controllers using STM32 microcontroller. To do that, we can also apply the Octave software for designing the PI controller, using phase margin calculations. Therefore, you should first initialize the buck converter parameters as the code below.

```
%clear all
pkg load control
s = tf('s');
%Initialization of converter parameters
%%%%%%%%%%%%%%%%%%%%%%%%%%%%%%%%%%%%%%%%%%%%%%%%
Vin = 24;
Vo = 16;
Io = 2.133;
Ro = Vo/Io;
fsw = 100000;
rL = 300e-3;
rESR = 3e-3;
Vds = 0.075;
Vfwd = 1.15;
iL = Io;
D = (Vfwd + iL*rL + Vo)/(Vin - Vds + Vfwd);
L = 200e-6;
C = 40e-6;
delta_iL = 0.1 * iL;
delta_Vo = 0.02 * Vo;
%L >= (Vin - Vds - iL*iL - Vo)*D/(fsw*delta_iL);
%C >= delta_iL/(8*delta_Vo*fsw);
%%%%%%%%%%%%%%%%%%%%%%%%%%%%%%%%%%%%%%%%%%%%%%%%
```

The parasitic values are inductor resistance, r_L, equivalent series resistance of capacitor, r_{ESR}, saturation voltage of MOSFET, V_{ds}, and forward voltage drop of diode, V_{fwd}, which have very little effect on output of the system, but it is useful to include them in practice. All of the commands in Octave can be executed in MATLAB as well. Then we initialize the crossover frequency which is very important in controller design. In power supply design, we mostly focus on crossover frequency of the converter frequency response. So, in power supply design, we are focusing on frequency-domain. If you choose the crossover frequency high, your converter will be very sensitive and it will respond basically to anything. So, we here do not focus on disturbances, however, we will mostly focus on robustness, so, if we change the set point, it will not affect the stability of the converter. We define crossover frequencies for current and voltage separately as given in the following code.

```
%Initialization of the crossover frequency
%%%%%%%%%%%%%%%%%%%%%%%%%%%%%%%%%%%%%%%%%%
wc_i = 2*pi*2000;
fc_i = wc_i/(2*pi);
wc_complex_i = wc_i*1i;
wc_v = 2*pi*2000;
fc_v = wc_v/(2*pi);
wc_complex_v = wc_v*1i;
targ_PM_deg_i = 59;
targ_PM_deg_v = 59;
%%%%%%%%%%%%%%%%%%%%%%%%%%%%%%%%%%%%%%%%%%
```

The codes for small signal output current to duty cycle and output voltage to duty cycle transfer functions are as below.

```
%Small Signal Transfer Functions
%%%%%%%%%%%%%%%%%%%%%%%%%%%%%%%%%%%%%%%%%%%%%%%%
%%%%%%%%%%%%%%%%%%%%%%%%%%%%%
%Output Current to Duty Cycle beginning
```

```
GIoD_num = (1+s*C*rESR)*(Vin - Vds + Vfwd);
GIoD_den = (L*C*Ro + L*C*rESR)*s^2 + (C*Ro*rL
+ C*rESR*iL + C*Ro*rESR + L)*s + rL + Ro;
GIoD = (GIoD_num/GIoD_den);
%Output Current to Duty Cycle ending
%Output Voltage to Duty Cycle beginning
GVoD_num = (1+s*C*rESR)*(Vin - Vds + Vfwd)*Ro;
GVoD_den = (L*C*Ro + L*C*rESR)*s^2 + (C*Ro*rL +
C*rESR*rL + C*Ro*rESR + L)*s + rL + Ro;
GVoD = (GVoD_num/GVoD_den);
%Output Voltage to Duty cycle ending
%%%%%%%%%%%%%%%%%%%%%%%%%%%%%%%%%%%%%%%%%%%%%%%%
%%%%%%%%%%%%%%%%%%%%%%%%%%%%
```

The realization of complex small signal output current to duty cycle and output voltage to duty cycle transfer functions at the crossover frequency is represented in the following code.

```
%Transfer functions to realize the gain and
phase at crossover frequency
%%%%%%%%%%%%%%%%%%%%%%%%%%%%%%%%%%%%%%%%%%%%%%%%
%%%%%%%%%%%%%%%%%%%%%%%%%%%%
%Output Current to Duty Cycle gain_phase
beginning
g = wc_complex_i;
GIoD_num_g = (1+g*C*rESR)*(Vin - Vds + Vfwd);
GIoD_den_g= (L*C*Ro + L*C*rESR)*g^2 + (C*Ro*rL
+ C*rESR*rL + C*Ro*rESR + L)*g + rL + Ro;
GIoD_g= (GIoD_num_g/GIoD_den_g);
%Output Current to Duty Cycle gain_phase ending
%Output Voltage to Duty Cycle gain_phase beginning
g = wc_complex_v;
GVoD_num_g = (1+g*C*rESR)*(Vin - Vds + Vfwd)*Ro;
GVoD_den_g = (L*C*Ro+ L*C*rESR)*g^2 + (C*Ro*rL +
C*rESR*iL + C*Ro*rESR + L)*g + rL + Ro;
GVoD_g = (GVoD_num_g/GVoD_den_g);
%Output Voltage to Duty Cycle gain_phase ending
```

```
%%%%%%%%%%%%%%%%%%%%%%%%%%%%%%%%%%%%%%%%%%%%%%%%
%%%%%%%%%%%%%%%%%%%%%%%%%%%%
```

The code required for current and voltage controller design is given as follows.

```
%Current Controller design begin
i_gain = abs(GIoD_g);
i_phase = angle(GIoD_g)*(180/pi);
targ_PM_rad_i = targ_PM_deg_i*(pi/180);
phase_comp_i = targ_PM_rad_i - (i_phase*(pi/180))
- pi;
fi_i = fc_i * tan(-1*phase_comp_i);
Kpi = (1/i_gain)/sqrt((fi_i/fc_i)^2 + 1);
wi_i = fi_i*2*pi;
%Current Controller design ends
%Voltage Controller design begin abs (GVOD_g);
v_gain = abs(GVoD_g);
v_phase = angle(GVoD_g)*(180/pi);
targ_PM_rad_v = targ_PM_deg_v*(pi/180);
phase_comp_v = targ_PM_rad_v - (v_phase* (pi/180))
- pi;
fi_v = fc_v * tan(-1*phase_comp_v);
Kpv = (1/v_gain)/sqrt((fi_v/fc_v)^2 + 1);
wi_v = fi_v*2*pi;
%Voltage Controller design ends
```

The codes for current and voltage controller transfer functions and also current and voltage closed-loop transfer functions are illustrated below.

```
%%%%%%%%%%%%%%%%%%%%%%%%%%%%%%%%%%%%%%%%%%%%%%%%
%%%%%%%%%%%%%%%%%%%%%%%%%%%%
Gci = Kpi*(wi_i/s + 1);
Gcv = Kpv* (wi_v/s + 1);
%Gc1 = 1;
Ti = Gci*GIoD;
Tv = Gcv*GVoD;
```

```
%Create a figure to hold the plots
%figure;
%Plot Bode plot for the system
%bode(Ti);
%bode(Tv);
%%%%%%%%%%%%%%%%%%%%%%%%%%%%%%%%%%%%%%%%%%%%%%%%
%%%%%%%%%%%%%%%%%%%%%%%%%%%%
```

The following codes are used for calculating gain and phase margins using closed-loop transfer functions.

```
%%%%%%%%%%%%%%%%%%%%%%%%%%%%%%%%%%%%%%%%%%%%%%%%
%%%%%%%%%%%%%%%%%%%%%%%%%%%%
%Calculate gain and phase margins
[gain_margin, phase_margin, crossover_
frequencies, phase_crossover] = margin(Ti);
%Display Current Bode Plot the results
fprintf('Current, Gain margin: %.2f dB\n',
20*log10 (gain_margin));
fprintf('Current Phase margin: %.2f degrees \n',
phase_margin);
fprintf('Current Crossover frequencies: %.2f
rad/s\n', crossover_frequencies);
fprintf('Current Phase at crossover: %.2f
degrees\n', phase_crossover);
[gain_margin, phase_margin, crossover_
frequencies, phase_crossover] = margin(Tv);
%Display Voltage Bode Plot the results
fprintf('Voltage Gain margin: %.2f dB\n',
20*log10 (gain_margin));
fprintf('Voltage Phase margin: %.2f degrees\n',
phase_margin);
fprintf('Voltage Crossover frequencies: %.2f
rad/s\n', crossover_frequencies);
fprintf('Voltage Phase at crossover: %.2f
degrees\n', phase_crossover);
%%%%%%%%%%%%%%%%%%%%%%%%%%%%%%%%%%%%%%%%%%%%%%%%
%%%%%%%%%%%%%%%%%%%%%%%%%%%%
```

For converting controller transfer functions from continuous to digital mode or from s-domain to z-domain, we use the sampling time of 100 μs as follows.

```
Gi_d = c2d(Gci, 100e-6);
Gv_d = c2d(Gcv, 100e-6);
```

So, in this case, the duty cycle is updated every 100 μs. We run the program and go to the command window and get current and voltage closed-loop gain and phase margins with current and voltage controllers with and without current and voltage controllers and then get the controllers transfer functions in z-domain as below.

```
>> loop_design_octave
Current, Gain margin: Inf dB
Current Phase margin: 59.06 degrees
Current Crossover frequencies: NaN rad/s
Current Phase at crossover: 12563.57 degrees
Voltage Gain margin: Inf dB
Voltage Phase margin: 58.94 degrees
Voltage Crossover frequencies: NaN rad/s
Voltage Phase at crossover: 12569.17 degrees
>> Gci
Transfer function 'Gci' from input 'u1' to output
...
    0.1594 s + 225.5
 y1: ----------------
           s
Continuous-time model.
>> Gcv
Transfer function 'Gcv' from input 'u1' to output
...
    0.02127 s + 30.16
 y1: -----------------
           s
Continuous-time model.
```

```
>> Gi_d
Transfer function 'Gi_d' from input 'u1' to
output ...
    0.1594 z - 0.1369
 y1:  -----------------
          z - 1
Sampling time: 0.0001 s
Discrete-time model.
>> Gv_d
Transfer function 'Gv_d' from input 'u1' to
output ...
    0.02127 z - 0.01825
 y1:  -------------------
          z - 1
Sampling time: 0.0001 s
Discrete-time model.
```

These z-domain transfer functions are actually what we want to use in our codes. The 'I_0' value in the initialization of converter parameters at the beginning does not affect so much on the designed robust system. So, the output current set point could be changed in the firmware as you like in the reasonable range and it will not make the system unstable. As mentioned before, we will use the PI controllers obtained by pidTuner tool of MATLAB; however, we can also use the PI controllers gained from phase margin calculations by Octave programming. The whole code in Octave software is as below.

```
%clear all
pkg load control
s = tf('s');
%Initialization of converter parameters
%%%%%%%%%%%%%%%%%%%%%%%%%%%%%%%%%%%%%%%%%%%%%%%%
Vin = 24;
Vo = 16;
Io = 2.133;
Ro = Vo/Io;
```

```
fsw = 100000;
rL = 300e-3;
rESR = 3e-3;
Vds = 0.075;
Vfwd = 1.15;
iL = Io;
D = (Vfwd + iL*rL + Vo)/(Vin - Vds + Vfwd);
L = 200e-6;
C = 40e-6;
delta_iL = 0.1 * iL;
delta_Vo = 0.02 * Vo;
%L >= (Vin - Vds - iL*iL - Vo)*D/(fsw*delta_iL);
%C >= delta_iL/(8*delta_Vo*fsw);
%%%%%%%%%%%%%%%%%%%%%%%%%%%%%%%%%%%%%%%%%%%%%%%
%Initialization of the crossover frequency
%%%%%%%%%%%%%%%%%%%%%%%%%%%%%%%%%%%%%%%%%
wc_i = 2*pi*2000;
fc_i = wc_i/(2*pi);
wc_complex_i = wc_i*1i;
wc_v = 2*pi*2000;
fc_v = wc_v/(2*pi);
wc_complex_v = wc_v*1i;
targ_PM_deg_i = 59;
targ_PM_deg_v = 59;
%%%%%%%%%%%%%%%%%%%%%%%%%%%%%%%%%%%%%%%%%
%Small Signal Transfer Functions
%%%%%%%%%%%%%%%%%%%%%%%%%%%%%%%%%%%%%%%%%%%%%%%
%%%%%%%%%%%%%%%%%%%%%%%%%%%%
%Output Current to Duty Cycle beginning
GIoD_num = (1+s*C*rESR)*(Vin - Vds + Vfwd);
GIoD_den = (L*C*Ro + L*C*rESR)*s^2 + (C*Ro*rL +
C*rESR*iL + C*Ro*rESR + L)*s + rL + Ro;
GIoD = (GIoD_num/GIoD_den);
%Output Current to Duty Cycle ending
%Output Voltage to Duty Cycle beginning
GVoD_num = (1+s*C*rESR)*(Vin - Vds + Vfwd)*Ro;
GVoD_den = (L*C*Ro + L*C*rESR)*s^2 + (C*Ro*rL +
C*rESR*rL + C*Ro*rESR + L)*s + rL + Ro;
```

```
GVoD = (GVoD_num/GVoD_den);
%Output Voltage to Duty cycle ending
%%%%%%%%%%%%%%%%%%%%%%%%%%%%%%%%%%%%%%%%%%%%%%%%
%%%%%%%%%%%%%%%%%%%%%%%%%%%%%
%Transfer functions to realize the gain and
phase at crossover frequency
%%%%%%%%%%%%%%%%%%%%%%%%%%%%%%%%%%%%%%%%%%%%%%%%
%%%%%%%%%%%%%%%%%%%%%%%%%%%%%
%Output Current to Duty Cycle gain_phase
beginning
g = wc_complex_i;
GIoD_num_g = (1+g*C*rESR)*(Vin - Vds + Vfwd);
GIoD_den_g = (L*C*Ro + L*C*rESR)*g^2 + (C*Ro*rL
+ C*rESR*rL + C*Ro*rESR + L)*g + rL + Ro;
GIoD_g = (GIoD_num_g/GIoD_den_g);
%Output Current to Duty Cycle gain_phase ending
%Output Voltage to Duty Cycle gain_phase
beginning
g = wc_complex_v;
GVoD_num_g = (1+g*C*rESR)*(Vin - Vds + Vfwd)*Ro;
GVoD_den_g = (L*C*Ro+ L*C*rESR)*g^2 + (C*Ro*rL +
C*rESR*iL + C*Ro*rESR + L)*g + rL + Ro;
GVoD_g = (GVoD_num_g/GVoD_den_g);
%Output Voltage to Duty Cycle gain_phase ending
%%%%%%%%%%%%%%%%%%%%%%%%%%%%%%%%%%%%%%%%%%%%%%%%
%%%%%%%%%%%%%%%%%%%%%%%%%%%%%
%Current Controller design begin
i_gain = abs(GIoD_g);
i_phase = angle(GIoD_g)*(180/pi);
targ_PM_rad_i = targ_PM_deg_i*(pi/180);
phase_comp_i = targ_PM_rad_i - (i_phase*(pi/180))
- pi;
fi_i = fc_i * tan(-1*phase_comp_i);
Kpi = (1/i_gain)/sqrt((fi_i/fc_i)^2 + 1);
wi_i = fi_i*2*pi;
%Current Controller design ends
%Voltage Controller design begin abs (GVOD_g);
v_gain = abs(GVoD_g);
```

```
v_phase = angle(GVoD_g)*(180/pi);
targ_PM_rad_v = targ_PM_deg_v*(pi/180);
phase_comp_v = targ_PM_rad_v - (v_phase*
(pi/180)) - pi;
fi_v = fc_v * tan(-1*phase_comp_v);
Kpv = (1/v_gain)/sqrt((fi_v/fc_v)^2 + 1);
wi_v = fi_v*2*pi;
%Voltage Controller design ends
%%%%%%%%%%%%%%%%%%%%%%%%%%%%%%%%%%%%%%%%%%%%%%%%
%%%%%%%%%%%%%%%%%%%%%%%%%%%%%
Gci = Kpi*(wi_i/s + 1);
Gcv = Kpv* (wi_v/s + 1);
%Gc1 = 1;
Ti = Gci*GIoD;
Tv = Gcv*GVoD;
%Create a figure to hold the plots
%figure;
%Plot Bode plot for the system
%bode(Ti);
%bode(Tv);
%%%%%%%%%%%%%%%%%%%%%%%%%%%%%%%%%%%%%%%%%%%%%%%%
%%%%%%%%%%%%%%%%%%%%%%%%%%%%%
%Calculate gain and phase margins
[gain_margin, phase_margin, crossover_
frequencies, phase_crossover] = margin(Ti);
%Display Current Bode Plot the results
fprintf('Current, Gain margin: %.2f dB\n',
20*log10 (gain_margin));
fprintf('Current Phase margin: %.2f degrees \n',
phase_margin);
fprintf('Current Crossover frequencies: %.2f
rad/s\n', crossover_frequencies);
fprintf('Current Phase at crossover: %.2f
degrees\n', phase_crossover);
[gain_margin, phase_margin, crossover_
frequencies, phase_crossover] = margin(Tv);
%Display Voltage Bode Plot the results
```

```
fprintf('Voltage Gain margin: %.2f dB\n',
20*log10 (gain_margin));
fprintf('Voltage Phase margin: %.2f degrees\n',
phase_margin);
fprintf('Voltage Crossover frequencies: %.2f
rad/s\n', crossover_frequencies);
fprintf('Voltage Phase at crossover: %.2f
degrees\n', phase_crossover);
%%%%%%%%%%%%%%%%%%%%%%%%%%%%%%%%%%%%%%%%%%%%%%%%%
%%%%%%%%%%%%%%%%%%%%%%%%%%%%%
Gi_d = c2d(Gci, 100e-6);
Gv_d = c2d(Gcv, 100e-6);
```

CHAPTER 3

Digital Control Implementation of the Buck Converter

3.1 HARDWARE OUTLINE

In this chapter, we are going to discuss about the hardware we will use to implement the buck converter controller. We will utilize the STM32 Nucleo-G474RE board and it is connected to the main converter board as shown in Figure 3.1.

We have actually an AC–DC converter that converts single phase 230 V_{ac} to 24 V_{dc} and that is an input to the system. In the main converter, we have an auxiliary power supply and, of course, we have an inductor coil, and on the right-hand side of it, we have a diode and on the left-hand side, there is a Metal-Oxide-Semiconductor Field-Effect Transistor (MOSFET) and the MOSFET is significantly over sized and it does not require a heat sink, but the diode requires a heat sink. Then we have the output of the main converter and the resistive load is connected to the output. There are four 10 μF parallel capacitors and also we have an inductor that serves outgoing current to the output and it

DOI: 10.1201/9781003541356-3

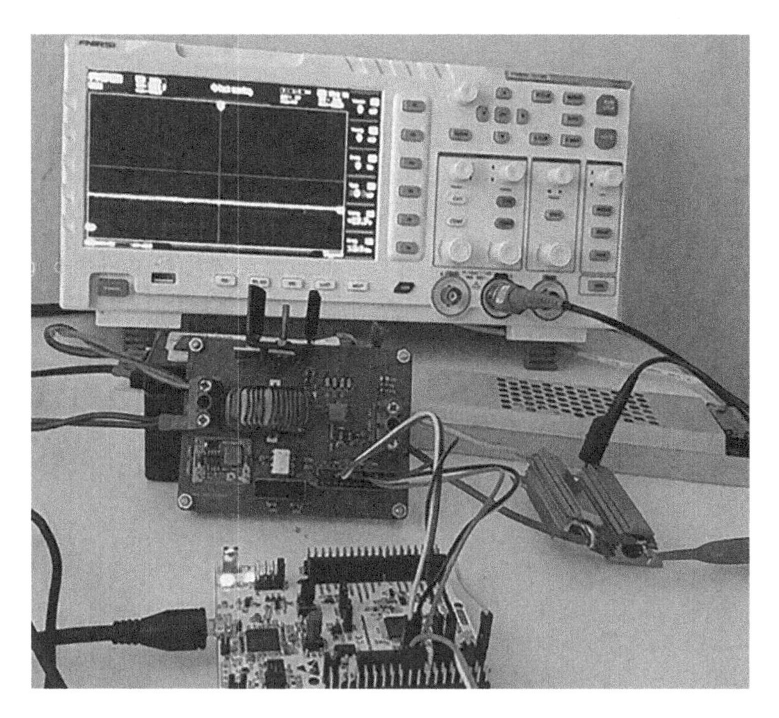

FIGURE 3.1 The buck converter controller implementation.

comes after the capacitors. Also, we have a shunt current-sensing resistor next to the inductor. There is another shunt current-sensing resistor of the main inductor. On the top right-hand side, we have a voltage sensor for the output. There is also a MOSFET driver in this case and we usually use bootstrap MOSFET driver for this particular application but we have not used it in this main converter board because they have some drawbacks, so, we use an isolated MOSFET driver. The converter is not considered isolated since the ground of input and output are the same. However, it helps go to any duty cycle we want. Next to the MOSFET driver, there is a power supply for it. It is a very common converter for this application. The STM32 Nucleo-G474RE board has high-resolution PWM and timer modules which are very useful in power electronics applications. The codes provided can be used

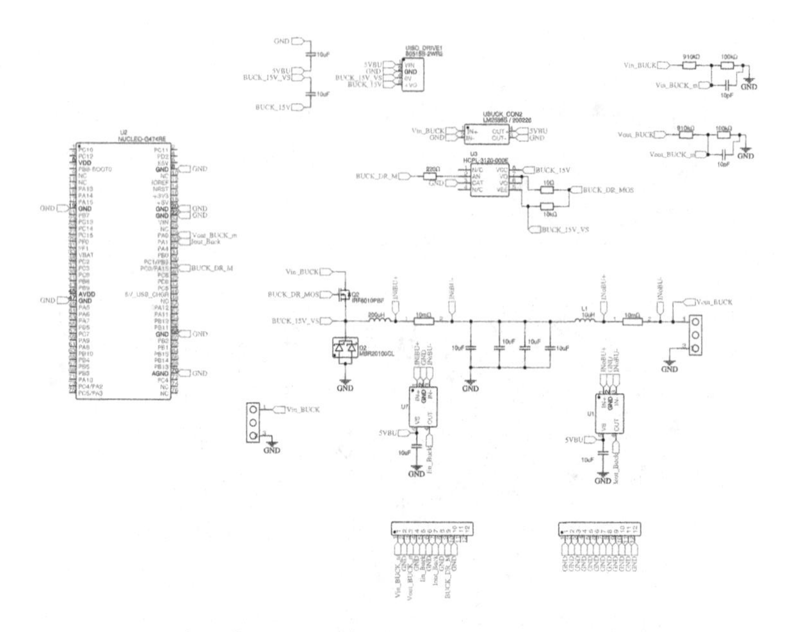

FIGURE 3.2 The schematic of buck converter and its controller.

on other STM32 microcontrollers as well. The buck converter and its controller schematic using the STM32 Nucleo-G474RE development board are depicted in Figure 3.2.

We have two current sensors and two voltage sensors in the schematic. The input voltage and current sensors are not used in this system. However, the output voltage and current sensors are utilized. We have also two input and output inductor current sensors and we can switch between them. We have used HCPL-3120 as the MOSFET driver. There is an auxiliary power supply, LM2596S, which generates 5 V at the output and its maximum input voltage is 28 V. There is also a power supply, B0515S, exclusively for the MOSFET driver which converts 5 V–15 V. The only deviation from traditional buck converter is that there is an output inductor, 10 μH, which is an outgoing for the next section of buck converter, for the load basically. We have also connections from buck converter board to the STM32 Nucleo-G474RE development board. The current sensor amplifier IC is INA180 from Texas Instruments Incorporated (TI).

3.2 FIRMWARE PERIPHERAL INITIALIZATION WITH CUBEMX

In this chapter, we are going to initialize the firmware peripherals to program the buck converter controller which is STM32 Nucleo-G474RE development board. To get started, you should note that the initialization is done using CubeMX software and it is done easily and already available with that. Therefore, we open the STM32CubeMX and under File tab, we create a New Project. In the Board Selector tab and in the Commercial Part Number, we do type NUCLEO-G474RE and select it in the bottom window and click on the Start Project button. In the popped-up window, we click on the Yes button to initialize all peripherals with their default mode to load the IOC and rendering UI. We select the Clock Configuration tab, the maximum clock frequency is 170 MHz. You can use pre-scaler to reduce the clock speed. However, in this case, we want to get maximum PWM resolution to realize the closed-loop control, so the maximum clock frequency is perfect. Then we go to the Project Manager tab, we do copy and paste the project directory in the Project Location also select a name, that is, BUCK_CONTROL in the Project Name. Then select STM32CubeIDE as Toolchain/IDE. We do not need to change anything in the Tools tab. Then we go back to the Pinout & Configuration tab. The STM32 microcontroller is capable of doing many tasks and we are going to initialize what is actually needed for the system. We start with the System Core and go to the DMA; however, before using DMA, the DMA is actually can be used in conjunction with the ADC and in the most convenient way in closed-loop controller is done by DMA. Then, in the ADC1 Mode and Configuration window, we set IN1 and IN2 as single-ended.

Then, we go back to the DMA and click on the Add button and select ADC1 item and choose Circular as Mode and Word as

Data Width. We do not discuss deeply any available peripherals, so we only consider peripherals as initialization perspective. Then we go back to the ADC1 Configuration window and in the Parameter Settings tab, we select independent mode for Mode, synchronous clock mode divided by 4 for Clock Pre-scaler, ADC 12-bit resolution, Data Alignment is not important, scan conversion mode is disabled, end of sequence conversion since we have two ADC channels, continuous conversion mode is enabled, and DMA continuous requests is enabled. In the ADC_Regular_ConversionMode part, we do enable regular conversions and oversampling is disabled. In the number of conversions, we choose 2 since we have 2 channels and for the sampling time of Channel 1, we select the maximum cycle, because it actually makes sure that the system is much stable so we sacrifice the sampling time to get much stability, so the Rank 1 is completed. We do the same settings for Rank 2 and Channel 2. Then we come back to the scan conversion mode in ADC_Settings part and you can see that is enabled now. We leave the other settings unchanged for ADC1. The next one is actually the timers. So, we go to the TIM1 in Timers section and this is specifically for PWM signal generation. We choose Internal Clock for Clock Source and select PWM Generation CH1 for Channel 1. You have not inversed output, so you select CH1 and not CH1N where its polarity is inversed relative to CH1 and you can see the TIM1_CH1 is actually in the pin PC0 in this case. In the Configuration window and Parameter Settings tab, we choose Counter Mode as Up and we select No Division in Internal Clock Division (CKD) for maximum resolution, also we do enable the auto-reload preload and you do not need to worry about the Break and Dead Time Management, it will play the role in the synchronous buck, boost or half-bridge converters and you need to consider the dead time management in those aspects. However, in this case, we will focus on Pre-scaler and Counter Period. The Counter Period and Pre-scaler are from 0 to 65535 ($2^{16} - 1$). The following equations can be used in this case:

$$F_{\text{PWM}} = \frac{F_{\text{CLK}}}{(\text{ARR}+1)\times(\text{PSC}+1)} \tag{3.1}$$

$$\text{ARR} = \frac{F_{\text{CLK}}}{(\text{PSC}+1)\times(F_{\text{PWM}})} - 1 \tag{3.2}$$

The Auto Reload Register (ARR) or Counter Period value is equivalent to the compare value and PSC is the Pre-scaler value. In this case, we assume the PWM frequency is 100 kHz and the clock frequency is 170 MHz. If, for example, the PSC is equal to 1, from Equation (3.2), we can calculate ARR which is 849. The next one is the timer interrupts, remember that in Chapter 2, to generate the digital controller coefficients, we selected the timer interrupts to be 100 μs which means its frequency is actually 10 kHz. We use TIM4 in this case to implement the timer interrupt. The first thing is to select Internal Clock as Clock Source and we will not have any outputs in this case, so, we will not select any channels. To do the parameter settings, we should note that the equation in this case is slightly different as below:

$$\text{ARR} = \frac{F_{\text{CLK}}}{(\text{PSC}+1)\times(F_{\text{INT}})} - 1 \tag{3.3}$$

The F_{INT} in this case is 10 kHz, and the same procedure done before is followed here. We select the PSC which is 1 in this case, also the F_{CLK} is 170 MHz, so the calculated ARR is to be 8499 and we keep the auto-reload preload to be Disable. So, everything is perfect and we can click on the Generate Code button. After generating the code, the STM32CubeIDE is automatically opened. So, the successful code generation message window is popped up finally, and then we click on the Open Project button. So, we launch the program in the workspace directory. Finally, you can see successfully importing the program into the workspace message. Then, we click on the project name to open it in STM32CubeIDE. We go to the main.c file in Src folder and see that the selected modules have been initialized. If you have forgotten to add some initialization modules, you can click on the BUCK_CONTROL.ioc file and in the popped up Open Associated Perspective window, click on the Yes button. Then, a new window called the Question window is popped up and we click on the No button to load the IOC. You may have mistaken and it is possible and that is basically what you need for initialization. In the next section, we will dive into coding the firmware and several other aspects that need to be taken into account.

3.3 MANUAL FIRMWARE INITIALIZATION

In this section, we are going to implement the second part of initialization of the firmware. This part is not done by CubeMX but it is very necessary for the functionality of the closed-loop control system. The first thing is the initialization of PWM and the timer actually we have done PWM with timer. In the previous section, we did timer interrupts on TIM4 and did the parameter settings, and then we go to the NVIC settings and enable the TIM4 global interrupt. We just save the changes and generate the code gain. Before doing anything, you should choose if the system giving you the PWM with the frequency that is actually required and, in this case, we have set it to be 100 kHz. The library we are using is the HAL library and it means Hardware Abstraction Layer and it is so convenient to use. Then, we go to start the PWM by writing the following code before the while (1) code:

```
HAL_TIM_PWM_Start(&htim1, TIM_CHANNEL_1);
```

Then, we are going to initialize the PWM duty cycle. We calculated PWM duty cycle to be between 0 and 849. We start with a duty cycle of 350 which is about 41.2% duty cycle by adding the following code inside the while (1) loop and run the program as depicted in Figure 3.3.

```
TIM1->CCR1 = 350;
```

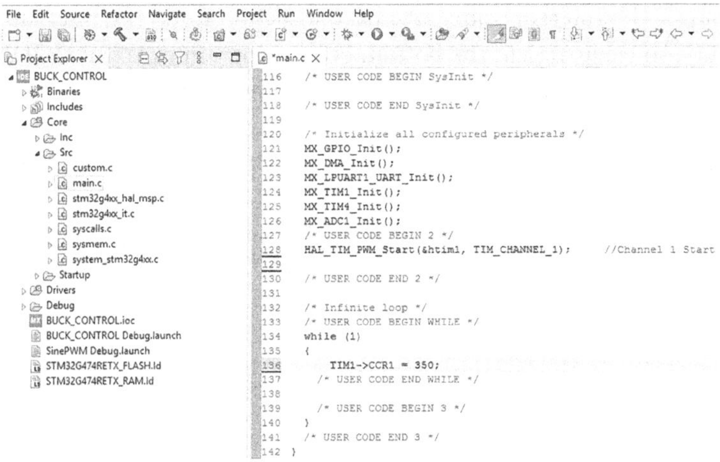

```
116    /* USER CODE BEGIN SysInit */
117
118    /* USER CODE END SysInit */
119
120    /* Initialize all configured peripherals */
121    MX_GPIO_Init();
122    MX_DMA_Init();
123    MX_LPUART1_UART_Init();
124    MX_TIM1_Init();
125    MX_TIM4_Init();
126    MX_ADC1_Init();
127    /* USER CODE BEGIN 2 */
128    HAL_TIM_PWM_Start(&htim1, TIM_CHANNEL_1);      //Channel 1 Start
129
130    /* USER CODE END 2 */
131
132    /* Infinite loop */
133    /* USER CODE BEGIN WHILE */
134    while (1)
135    {
136        TIM1->CCR1 = 350;
137      /* USER CODE END WHILE */
138
139      /* USER CODE BEGIN 3 */
140    }
141    /* USER CODE END 3 */
142  }
```

FIGURE 3.3 Adding code to start the PWM with a duty cycle of 41.2%.

If you connect pin PC0 to the oscilloscope, you can see the PWM signal with a frequency of 100 kHz and a duty cycle of 41.2%. We will use a maximum of 70% of duty cycle in this program. Next, we are going to initialize the ADC and DMA which is actually done together in this case. As a first thing, we will utilize the ADC with DMA and add the following code before while (1):

```
HAL_ADC_Start_DMA(&hadc1, (uint32_t *)
adcValues, 2);
Also, in the Private variables section, you need
to set up the code below.
float Vactual = 0.0;
float Iactual = 0.0;
uint32_t adcValues[2];
```

It is better to initialize the timer interrupt as well so we add the following code before while (1).

```
HAL_TIM_Base_Start_IT(&htim4);
```

That code actually enables the timer interrupts to get started, the TIM4 counter is counting up and when it is overflowed it can trigger the interrupt duty. We need a section to do the interrupt duty, so we just go to the bottom of the program in USER CODE 4 section and add the following code for interrupt duty.

```
void HAL_TIM_PeriodElapsedCallback(TIM_
HandleTypeDef* htim){
     if(htim->Instance == TIM4){
           Vactual = adcValues[0]*0.008137207031;
           Iactual = adcValues[1]*0.001611328125;
     }
}
```

Once, initializing the timer interrupts, it starts to counting up at the beginning of initialization and when it overflows it executes the above code and then checks the Instance, if it is TIM4, then it is triggered and then it executes the codes inside of it. As you know, we have used resistor dividers to sense input or output voltages, there are applications such as solar or battery charge controller, we are interested to sense the input voltage; however, in this case, we have used output voltage sensing for our system. The following equations are used to calculate the V_{actual}:

$$V_{\text{actual}} = \left(\left(\left(\frac{\text{RawAdcVal}}{2^{\text{AdcRes}}}\right) \times V_{\text{ref}}\right) \times \frac{R_1 + R_2}{R_2}\right) \tag{3.4}$$

$$V_{\text{actual}} = \left(\left(\left(\frac{\text{RawAdcVal}}{4096}\right) \times 3.3\right) \times \frac{910\text{k} + 100\text{k}}{100\text{k}}\right) \tag{3.5}$$

$$V_{\text{actual}} = \text{RawAdcVal} \times \frac{V_{\text{ref}}}{2^{\text{AdcRes}}} \times \frac{R_1 + R_2}{R_2} \tag{3.6}$$

$$V_{\text{actual}} = \text{RawAdcVal} \times \frac{3.3}{4096} \times \frac{910\text{k} + 100\text{k}}{100\text{k}} \tag{3.7}$$

$$V_{\text{actual}} = \text{RawAdcVal} \times 0.008137207031 \tag{3.8}$$

For the I_{actual}, we have something similar:

$$I_{\text{actual}} = \frac{\left(\frac{\left(\frac{\text{RawAdcVal}}{2^{\text{AdcRes}}}\right) \times V_{\text{ref}}}{\text{AmpGain}}\right)}{R_{\text{shunt}}} \tag{3.9}$$

$$I_{\text{actual}} = \frac{\left(\frac{\left(\frac{\text{RawAdcVal}}{4096}\right) \times 3.3}{50}\right)}{0.01} \tag{3.10}$$

You should note that the voltage read in the STM32 microcontroller ADC pin is as below:

$$V_{\text{read}} = \left(\frac{\text{RawAdcVal}}{4096}\right) \times 3.3 \tag{3.11}$$

For calculating the actual voltage across the shunt resistor, we should divide the voltage read in STM32 microcontroller ADC pin by the amplifier gain, AmpGain, which is for INA180 differential current sense amplifier is 50. Then for obtaining the I_{actual}, we should divide the actual voltage across the shunt resistor by the shunt resistor value, which is 10 mΩ as you can see in the following equations:

$$I_{\text{actual}} = \text{RawAdcVal} \times \frac{\left(\frac{\left(\frac{3.3}{4096}\right)}{50}\right)}{0.01} \tag{3.12}$$

$$I_{\text{actual}} = \text{RawAdcVal} \times 0.001611328125 \tag{3.13}$$

Then we can run the program and debug the code. So, we run the code in the debug mode to verify that the V_{actual} and I_{actual} are really updated with the output voltage and output current. That is very important before going ahead to control the converter and first thing is to check that the duty cycle can be changed and the second part is to check the timer interrupt is being triggered and it is actually very important to execute the control code and the third part is ensuring that you can read the actual output voltage and current with the equations have been proposed because they are very important to obtaining closed-loop control. If one of them is not set, nothing will be worked basically. We have checked the PWM signal before but the others have not been checked yet and they will be demonstrated in this section. Therefore, for running the program in debug mode, we do right click on the project name, select Debug As and then choose Debug Configurations… item. In the Debug Configurations window and in the Debugger tab, we set the Enable live expressions and then click on the Close button.

Then, we click on the Debug icon (green insect). In the opened Confirm Perspective Switch window, we do click on the Switch button. From Windows tab in menu, we select Show View and the Expressions item. Then we click on the Add new expression, to see the expressions we want, V_{actual} is one of them and I_{actual} is being the other one and we want to show up them, so, we click on the Resume (F8) icon in menu. The duty cycle of the PWM signal has been set to as below equation:

$$\text{PWM duty cycle} = \frac{400}{849} \times 100 \approx 47\% \tag{3.14}$$

The system is turned on and the values of V_{actual} and I_{actual} are displayed as Table 3.1.

TABLE 3.1 Displayed V_{actual} and I_{actual} Values

Expression	Type	Value
V_{actual}	float	10.8957205
I_{actual}	float	1.44052732

We see that values are not stable and changing because it is an open-loop control system and we have not added the controller codes to make a closed-loop controller yet, so, that is because it is uncompensated or open-loop control system. In this case, the output resistor is about 7.5 Ω. Now, we connect the output voltage to the oscilloscope to display the output voltage on it, we can see that the value on the oscilloscope screen is different from the value displayed on live expressions in debug mode running. Because the oscilloscope used in this case has not been calibrated that well, so, there is a slight difference with what it senses so it reads 12.4 volts but in actuality when you use a multimeter, you will see more accurate result of 11.48. We did manual initialization of our code and used the live expressions in debug mode running the program. In the next section, we will implement the voltage and current closed-loop controller and will add them to our code.

3.4 FIRMWARE IMPLEMENTATION OF VOLTAGE CONTROL

In this section, we will discuss about the next part of firmware which is implementing the closed-loop control system, especially the function that will implement the closed-loop control system, that is, changing the duty cycle based on the center inputs. We will think about the broader scheme which in terms of codes in it and code the firmware in such a way that be easily used in other families or use a function or separate file, that you actually call to implement the controller, so that it can be used in other microcontrollers as well and you basically copy and paste certain file and change the duty cycle update and also do ADC update from the DMA. The first step is that basically, you should create a header file, so we do right click on the Inc folder, select New, and then choose the Header File option. We do type a name, custom.h, in Header file row and click on the Finish button. We define a void function inside custom.h file as below, since it will not return anything and save it.

```
#ifndef INC_CUSTOM_H_
#define INC_CUSTOM_H_
void CNTRL_ROUTINE(void);
#endif
```

After that, we do right click on the Src folder, select New, and then choose the Source File option. We do type a name, custom.c, in Source file row and click on the Finish button. First, we are going to make the function we defined before, so we add the following code in custom.c file.

```
void CNTRL_ROUTINE(void){
}
```

A lot of extern variables need to be defined inside that function, so, we are going to add them from main.c file, in the Private variables (PV) section, we add two variables, V_{ref} and I_{ref}, then you will need to set up arrays to store the variables for the pole zero compensator or digital controller so we define voltage and current arrays for errors, e_v and e_i, also we define voltage and current arrays for outputs, y_v and y_i, and initialize them to zero as below.

```
/* USER CODE BEGIN PV */
float Vref = 16.0;
float Iref = 2.133;
float Vactual = 0.0;
float Iactual = 0.0;
float e_v[5] = {0};
float y_v[5] = {0};
```

```
float e_i[5]  = {0};
float y_i[5]  = {0};
uint32_t adcValues[2];
/* USER CODE END PV */
```

Remembering from Equation (2.30), we have substituted e_v instead of e, and e_m instead of y, so e_n = e_v[0], e_{n-1} = e_v[1], etc., and also y_n = y_v[0], y_{n-1} = y_v[1], etc., and y_n is actually the duty cycle update. The closed-loop block diagram of the system is illustrated in Figure 3.4.

Then, we go to the custom.c file and add them as extern variables as given in the following code.

```
#include <stdio.h>
#include <stdint.h>
#include <stdlib.h>
#include main.h
#include custom.h
extern float Vref;
extern float Iref;
extern float Vactual;
extern float Iactual;
extern float e_v[5];
extern float y_v[5];
extern float e_i[5];
extern float y_i[5];
extern uint32_t adcValues[2];
void CNTRL_ROUTINE(void){
}
```

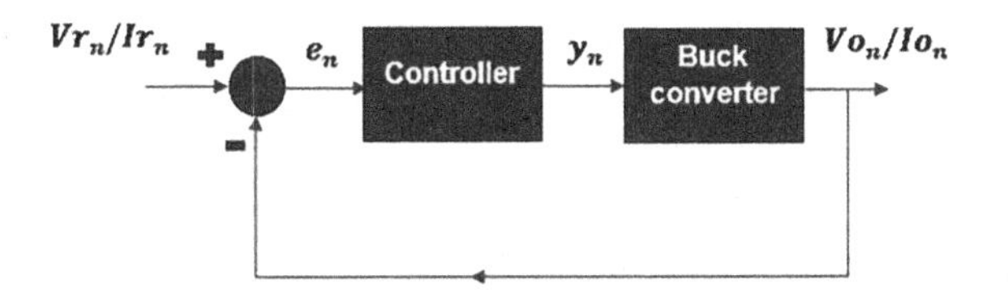

FIGURE 3.4 The closed-loop block diagram of the system.

Then, we should add necessary libraries to the custom.c file as you can see in the above code; however, including math.h library is not necessary in this case. So, we are going inside the CNTRL_ROUTINE function. From Equation (2.30), we should note that the y_n(y[0]) is used to update the duty cycle and y_{n-1}(y[1]) is the history 1, y_{n-2}(y[2]) is the history 2, and y_{n-3}(y[3]) is the history 3. After implementing them, you need to update each one, so y_{n-1} will be equated to what y_n is, y_{n-2} will be equated to what y_{n-1} is and y_{n-3} will be equated to what y_{n-2} is as below:

$$y_{n-1} \rightarrow y_n \tag{3.15}$$

$$y_{n-2} \rightarrow y_{n-1} \tag{3.16}$$

$$y_{n-3} \rightarrow y_{n-2} \tag{3.17}$$

The same updates are applied for the errors, so, first, we do copy and paste the V_{actual} equation code from interrupt duty function inside the CNTRL_ROUTINE function and apply the updates for outputs and errors as depicted in the following code.

```
void CNTRL_ROUTINE(void){
  Vactual = adcValues[0]*0.008137207031;
  y_v[3] = y_v[2];
  y_v[2] = y_v[1];
  y_v[1] = y_v[0];
  e_v[3] = e_v[2];
  e_v[2] = e_v[1];
  e_v[1] = e_v[0];
  e_v[0] = Vref - Vactual;
  y_v[0] = A1_v*y_v[1] + A2_v*y_v[2] +
  A3_v*y_v[3] + B0_v*e_v[0] + B1_v*e_v[1] +
  B2_v*e_v[2] + B3_v*e_v[3];
```

```
    if(y_v[0] > y_v_max){
        y_v[0] = y_v_max;
    }
    if(y_v[0] < y_v_min){
        y_v[0] = y_v_min;
    }
    TIM1->CCR1 = (int)(MAX_DUTY*y_v[0]);
}
```

Then we add the codes for e_v[0] and PI controller and define limits for maximum and minimum duty cycles, y_v[0]. Since the y_v[0] is of float type and is between 0 and 1, we should convert it to integer and we know that the MAX_DUTY is 849 in this case as you can see in the above code. It is the great advantage of STM32 microcontrollers that you can implement a controller mathematically and allow you to define float-type numbers and it usually does the mathematical operations in less than 10 μs. Now we are going to define coefficients and variables that have been utilized in the code inside the custom.c file as the following code.

```
#define        A1_v           1
#define        A2_v           0
#define        A3_v           0
#define        B0_v           0.001107
#define        B1_v           0.002056
#define        B2_v           0
#define        B3_v           0
#define        y_v_max        0.75
#define        y_v_min        0.0
#define        MAX_DUTY       849
```

In terms of protecting your microcontroller, one thing should be considered and it is the constant duty cycle in Equation (2.1), so we should note that when setting the maximum value of duty cycle. Therefore, we can choose y_v_max to be a little bit more than D value, that is, 0.75 as you can see in the above code. So, it will not go above 75% duty cycle. Next, we should call the function CNTRL_ROUTINE, so we go back to the main.c file and call it inside the interrupt duty function and do comment the V_{actual} and I_{actual} expressions, as depicted in the following code, since they will be calculated inside the CNTRL_ROUTINE function in custom.c file.

```
/* USER CODE BEGIN 4 */
void HAL_TIM_PeriodElapsedCallback(TIM_
HandleTypeDef* htim){
      if(htim->Instance == TIM4){
            CNTRL_ROUTINE();
            // Vactual = adcValues[0]
            *0.008137207031;
            // Iactual = adcValues[1]
            *0.001611328125;
      }
}
/* USER CODE END 4 */
```

Before starting to build the program, we should remove the following code inside the loop while (1).

```
TIM1->CCR1=400;
```

Then, we do build the program by clicking on the Build All (Ctrl+B) icon in the menu. The program is successfully built and we can now run it by clicking on the run icon in the menu. So, the code is successfully uploaded to the STM32 microcontroller, now we can see the output voltage using the oscilloscope and multimeter. As you can see on the oscilloscope, the voltage displayed is 17.9 V; however, on the multimeter screen, the voltage value is 16.36 V. You should note that the V_{ref} has been set to 16 V. In the next section, we will evaluate the system operation for different set points.

3.5 FIRMWARE IMPLEMENTATION OF CURRENT CONTROL

In this section, we will discuss about implementation of current closed-loop control specifically output current to duty cycle which will be used to obtain the coefficients of the closed-loop controller and used for controlling the output current of buck converter. Before we actually start, we are going to change the setpoint to get the goal of robustness, so we change the set point V_{ref} to 10 V and see the results to ensure the operation of implemented voltage closed-loop control. Therefore, in the main.c file, we modify the code as below, and then run the program.

```
/* USER CODE BEGIN PV */
float Vref = 10.0;
float Iref = 2.133;
```

As you can see, the voltage value on the oscilloscope is 11.5 V; however, the output voltage on the multimeter screen is 10.41 V which is much less than 11.5 V. As you can see, there is a slight offset from what we actually set up. In essence of how to deal with this is basically to increase the average error, the average voltage error is to use a more precise voltage sensor that will also affect the loop and you will incorporate it with the loop design. Another thing you can do is that you simply adjust the reference voltage, V_{ref}, and make it slightly lower and cause the offset value to change. In many applications that are actually utilized, you do want to be in a certain range, for example, if you set the output voltage to be 3.3 V, and you want to get about 3.3 V–3.5 V and you want it to be in certain range. You should know there are many factors to cause the output voltage to be exactly 10 V. Now, we are going to implement the output current closed-loop control, so, first of all in the custom.h file, we add the new function, CNTRL_ROUTINE_CURRENT, and change the name of the voltage control function, CNTRL_ROUTINE, to CNTRL_ROUTINE_VOLTAGE as shown in the following code.

```
#ifndef INC_CUSTOM_H_
#define INC_CUSTOM_H_
void CNTRL_ROUTINE_VOLTAGE(void);
void CNTRL_ROUTINE_CURRENT(void);
#endif
```

Then we are going to the file custom.c and we create the function CNTRL_ROUTINE_CURRENT, exactly like the function CNTRL_ROUTINE_VOLTAGE, and we use the I_{actual} code line instead of V_{actual} in this case, and also change any *v*, in arrays or variables to *i*, as you can see in the code below.

```
void CNTRL_ROUTINE_CURRENT(void){
  Iactual = adcValues[1]*0.001611328125;
  y_i[3] = y_i[2];
  y_i[2] = y_i[1];
  y_i[1] = y_i[0];
  e_i[3] = e_i[2];
  e_i[2] = e_i[1];
  e_i[1] = e_i[0];
  e_i[0] = Iref - Iactual;
  y_i[0] = A1_i*y_i[1] + A2_i*y_i[2] +
  A3_i*y_i[3] + B0_i*e_i[0] + B1_i*e_i[1] +
  B2_i*e_i[2] + B3_i*e_i[3];
  if(y_i[0] > y_i_max){
     y_i[0] = y_i_max;
  }
  if(y_i[0] < y_i_min){
     y_i[0] = y_i_min;
  }
  TIM1->CCR1 = (int)(MAX_DUTY*y_i[0]);
}
```

We know that the output resistor value is 7.5 Ω and we set the I_{actual} to 2.133 A, so, the output voltage will be 16 V. We have used 5% resistors for resistor divider sensors to get more accurate output voltage it is better to use 1% resistors to achieve more accurate values. Then we are going to define the PI current closed-loop controller coefficients as well as maximum and minimum values for the duty cycles as depicted in the following code.

```
#define        A1_v           1
#define        A2_v           0
```

```
#define        A3_v          0
#define        B0_v          0.001107
#define        B1_v          0.002056
#define        B2_v          0
#define        B3_v          0
#define        A1_i          1
#define        A2_i          0
#define        A3_i          0
#define        B0_i          0.008334
#define        B1_i          0.01548
#define        B2_i          0
#define        B3_i          0
#define        y_i_max       0.75
#define        y_i_min       0.0
#define        y_v_max       0.75
#define        y_v_min       0.0
#define        MAX_DUTY      849
```

Then we go to the interrupt duty function and call the current controller function inside of it as given in the code below.

```
/* USER CODE BEGIN 4 */
void HAL_TIM_PeriodElapsedCallback(TIM_
HandleTypeDef* htim){
     if(htim->Instance == TIM4){
        // CNTRL_ROUTINE_VOLTAGE();
           CNTRL_ROUTINE_CURRENT();
       }
}
/* USER CODE END 4 */
```

You should note that we do not call the voltage controller function here, so we do comment it. Then we can build the program by clicking on the Build All (Ctrl+B) icon in the menu and it is built successfully. Since, the I_{actual} is 2.133 A, and the output resistor is 7.5 Ω, we should expect to have the output voltage in the range from 16 V to 16.5 V. So, we can run the program to upload it to the STM32 microcontroller and the download is verified successfully.

Now, we can see the output voltage value on the oscilloscope which is 18.1 V; however, it is 16.51 V on the multimeter screen. That is all about the current closed-loop controller, and basically all of the works have been done. We want to have precise current closed-loop controller, so we used a shunt in this case. However, when you are going to sense high currents above 15 A, it is not advisable to use a shunt because the shunt color is changing and also its resistance is changing at some point and affects the efficiency of the system. In certain cases, you want to utilize Hall effect current sensors and they are much better and they are highly efficient as well and also its resistance is almost negligible. So, in that case, Equation (3.13) for I_{actual} will not suffice. How do we modify the equation in case when we use the Hall effect sensor? The Hall effect sensor essentially works with offset. It is usually an offset value and sensitivity. For example, the ACS712 current sensor is a Hall effect current sensor and it has an offset value because it has been designed to work with DC and AC systems and we should scale the raw ADC value with the offset value and we should subtract offset from it however we prefer to subtract it from offset value since we will not pass 2.5 V in this case since we assume the current is flowing in reverse direction, that is, from IP- to IP+ in ACS712 (–20 A to 20 A). So, we have the following equations:

$$I_{\text{actual}} = \left(\frac{\text{OffsetVal} - \left(\dfrac{\text{RawAdcVal}}{2^{\text{AdcRes}}} \times V_{\text{ref}} \right)}{\text{Sensitivity}} \right) \tag{3.18}$$

$$I_{\text{actual}} = \frac{\text{OffsetVal}}{\text{Sensitivity}} - \frac{\text{RawAdcVal} \times V_{\text{ref}}}{\text{Sensitivity} \times 2^{\text{AdcRes}}} \tag{3.19}$$

$$\text{OffsetVal} = 2.5\,\text{V} \tag{3.20}$$

$$\text{Sensitivity} = 0.1\,\text{V}/\text{A} \tag{3.21}$$

$$V_{\text{ref}} = 3.3\,\text{V} \tag{3.22}$$

$$2^{\text{AdcRes}} = 2^{12} = 4096 \tag{3.23}$$

$$I_{\text{actual}} = 25 - \text{RawAdcVal} \times 0.008056640625 \tag{3.24}$$

So, we can use it in the function CNTRL_ROUTINE_CURRENT in custom.c file as given in the following code.

```
void CNTRL_ROUTINE_CURRENT(void){
// Iactual = adcValues[1]*0.001611328125;
     Iactual = 25-adcValues[1]*0.008056640625;
```

You should note that the Hall effect current sensors are very prone to be affected by magnetic fields. In the next section, we will discuss about average current mode control.

3.6 AVERAGE CURRENT MODE CONTROL OF THE BUCK CONVERTER

In this section, we will deal with average current mode control and essentially what it means you actually control the current through the inductor, so instead of controlling the output voltage or output current, you basically control the current flowing through the inductor and what that tells us, you have double loop also it is called double loop control, and what you get in double loop control is that the outer voltage loop makes a reference for the inner current loop and it really presents many advantages in the converter control because you can essentially ensure that your converter constantly operates in continuous conduction mode, CCM, and additionally you can actually save your MOSFET from a lot of stress because inherently you are limited in the amount of the current goes through your inductor by proxy MOSFET or diode as well, essentially the amount of current goes through the converter so you basically limit it in every cycle as opposed you have not checked it in other methods, that is, the output voltage to duty cycle or output current to duty cycle where you control them like a single loop control. If you want to limit current, you essentially have to use other than the duty cycle method, we keep our fingers crossed, or, literally, you have to use another if statement, that is, if the current above the limit reduces the duty cycle as opposed to double control loop inherently and you just limit the current by default as you are using the ADC of course. In this section, we will dive into practical implementation of this method and we will not go through the theoretical analysis because the average current mode control in theoretical is rather complicated and we will go to other approach so we will implement the average current mode in the firmware and we are going to start coding it. You should note that the outer control loop is quite slow, so, every time the outer control loop is executed, the current control loop is executed 10 times, so the voltage control loop only executes 1 iteration for 10 iterations of current loop control. Therefore, in main.c file and Private variables section, we define a loop counter variable (loop_cnt) and initialize it to zero as given in the code below.

```
/* USER CODE BEGIN PV */
float Vref = 10.0;
float Iref = 1.2;
float Vactual = 0.0;
float Iactual = 0.0;
float e_v[5] = {0};
float y_v[5] = {0};
float e_i[5] = {0};
float y_i[5] = {0};
int loop_cnt = 0;
uint32_t adcValues[2];
/* USER CODE END PV */
```

Then, we go to the custom.h file and define the functions we require in this case as you can see in the following code.

```
#ifndef INC_CUSTOM_H_
#define INC_CUSTOM_H_
void CNTRL_ROUTINE_VOLTAGE(void);
void CNTRL_ROUTINE_CURRENT(void);
void DOUB_LOOP(void);
void DOUB_LOOP_VOLTAGE(void);
void DOUB_LOOP_CURRENT(void);
#endif
```

You should note that the first function is for the entire double loop controller, the second one is for the voltage control loop, and the third one is for the current control loop.

In custom.c file, we define the loop counter variable (loop_cnt) as external integer and add the codes inside the entire double loop control function and call the double loop current control in every iteration and double loop voltage control in every 10 iterations as given in the following code.

```
#include <stdio.h>
#include <stdint.h>
#include <stdlib.h>
#include main.h
#include custom.h
#define       A1_v          1
#define       A2_v          0
#define       A3_v          0
#define       B0_v          0.001107
#define       B1_v          0.002056
#define       B2_v          0
#define       B3_v          0
#define       A1_i          1
#define       A2_i          0
#define       A3_i          0
#define       B0_i          0.008334
```

```
#define         B1_i            0.01548
#define         B2_i            0
#define         B3_i            0
#define         y_i_max         0.75
#define         y_i_min         0.0
#define         y_v_max         0.75
#define         y_v_min         0.0
#define         MAX_DUTY        849
extern float Vref;
extern float Iref;
extern float Vactual;
extern float Iactual;
extern float e_v[5];
extern float y_v[5];
extern float e_i[5];
extern float y_i[5];
extern uint32_t adcValues[2];
extern int loop_cnt;
void CNTRL_ROUTINE_VOLTAGE(void){
  Vactual = adcValues[0]*0.008137207031;
  y_v[3] = y_v[2];
  y_v[2] = y_v[1];
  y_v[1] = y_v[0];
  e_v[3] = e_v[2];
  e_v[2] = e_v[1];
  e_v[1] = e_v[0];
  e_v[0] = Vref - Vactual;
  y_v[0] = A1_v*y_v[1] + A2_v*y_v[2] +
  A3_v*y_v[3] + B0_v*e_v[0] + B1_v*e_v[1] +
  B2_v*e_v[2] + B3_v*e_v[3];
  if(y_v[0] > y_v_max){
     y_v[0] = y_v_max;
  }
  else if(y_v[0] < y_v_min){
     y_v[0] = y_v_min;
  }
  TIM1->CCR1 = (int)(MAX_DUTY*y_v[0]);
}
```

```
void CNTRL_ROUTINE_CURRENT(void){
  Iactual = adcValues[1]*0.001611328125;
  y_i[3] = y_i[2];
  y_i[2] = y_i[1];
  y_i[1] = y_i[0];
  e_i[3] = e_i[2];
  e_i[2] = e_i[1];
  e_i[1] = e_i[0];
  e_i[0] = Iref - Iactual;
  y_i[0] = A1_i*y_i[1] + A2_i*y_i[2] +
  A3_i*y_i[3] + B0_i*e_i[0] + B1_i*e_i[1] +
  B2_i*e_i[2] + B3_i*e_i[3];
  if(y_i[0] > y_i_max){
     y_i[0] = y_i_max;
  }
  else if(y_i[0] < y_i_min){
     y_i[0] = y_i_min;
  }
  TIM1->CCR1 = (int)(MAX_DUTY*y_i[0]);
}
void DOUB_LOOP(void){
  loop_cnt++;
  if(loop_cnt == 10){
       loop_cnt = 0;
       DOUB_LOOP_VOLTAGE();
  }
  DOUB_LOOP_CURRENT();
}
void DOUB_LOOP_VOLTAGE(void);
void DOUB_LOOP_CURRENT(void);
```

You should note that the double loop current control function, DOUB_LOOP_CURRENT, is identical to the current control routine function, CNTRL_ROUTINE_CURRENT, however, the double loop voltage control function, DOUB_LOOP_VOLTAGE, is slightly different from the voltage control routine function, CNTRL_ROUTINE_VOLTAGE. We do copy and paste the codes inside the CNTRL_ROUTINE_VOLTAGE function to inside the

DOUB_LOOP_VOLTAGE function and remove the last code line of it since the output of this function is the reference current, I_{ref}, and it will be used as a reference in the DOUB_LOOP_CURRENT function, so we add the reference current code as given in the code below.

```
void DOUB_LOOP_VOLTAGE(void){
  Vactual = adcValues[0]*0.008137207031;
  y_v[3] = y_v[2];
  y_v[2] = y_v[1];
  y_v[1] = y_v[0];
  e_v[3] = e_v[2];
  e_v[2] = e_v[1];
  e_v[1] = e_v[0];
  e_v[0] = Vref - Vactual;
  y_v[0] = A1_v*y_v[1] + A2_v*y_v[2] +
  A3_v*y_v[3] + B0_v*e_v[0] + B1_v*e_v[1] +
  B2_v*e_v[2] + B3_v*e_v[3];
  if(y_v[0] > y_v_max){
     y_v[0] = y_v_max;
  }
  else if(y_v[0] < y_v_min){
     y_v[0] = y_v_min;
  }
  Iref = Iref_max*y_v[0];
}
void DOUB_LOOP_CURRENT(void);
```

The Iref_max is the maximum current which is essentially required and that is the maximum inductor current you want, and we never get the inductor current above that value. Then we go back to the custom.c file and define Iref_max as 2.133 and change y_v_max from 0.75 to 1.0 as you can see in the following code.

```
//#define        y_v_max           0.75
#define          y_v_max           1.0
```

```
#define        y_v_min        0.0
#define        Iref_max       2.133
```

Then we do copy and paste the codes inside the CNTRL_ROUTINE_CURRENT function to inside the DOUB_LOOP_CURRENT function and you have used I_{ref} which has been obtained directly from the double loop voltage control function, DOUB_LOOP_VOLTAGE, as given in the following code.

```
void DOUB_LOOP_CURRENT(void){
  Iactual = adcValues[1]*0.001611328125;
  y_i[3] = y_i[2];
  y_i[2] = y_i[1];
  y_i[1] = y_i[0];
  e_i[3] = e_i[2];
  e_i[2] = e_i[1];
  e_i[1] = e_i[0];
  e_i[0] = Iref - Iactual;
  y_i[0] = A1_i*y_i[1] + A2_i*y_i[2] +
  A3_i*y_i[3] + B0_i*e_i[0] + B1_i*e_i[1] +
  B2_i*e_i[2] + B3_i*e_i[3];
  if(y_i[0] > y_i_max){
     y_i[0] = y_i_max;
  }
  else if(y_i[0] < y_i_min){
     y_i[0] = y_i_min;
  }
  TIM1->CCR1 = (int)(MAX_DUTY*y_i[0]);
}
```

So, the reference current, I_{ref}, basically keeps the inductor current in CCM, which is literally constant, remember that I_{ref} is running for 10 cycles and the function DOUB_LOOP_CURRENT runs for 10 times at first before start up the function DOUB_LOOP_VOLTAGE, so you should note that adequate I_{ref} should be there and not be too high that is why we initiate it to 0.2 A first, until it gets updated inside the function DOUB_LOOP_VOLTAGE, so we do the initialization of I_{ref} inside the main.c file as given in the code below.

```
/* USER CODE BEGIN PV */
float Vref = 10.0;
float Iref = 0.2;
```

Also, as we mentioned earlier, the function we want to call is the DOUB_LOOP function in every 100 μs inside the interrupt duty function. Therefore, we go to the main.c file and find the interrupt duty function and call the DOUB_LOOP function inside of it as you can see in the following code.

```
/* USER CODE BEGIN 4 */
void HAL_TIM_PeriodElapsedCallback(TIM_
HandleTypeDef* htim){
     if(htim->Instance == TIM4){
        DOUB_LOOP();
        // CNTRL_ROUTINE_VOLTAGE();
        // CNTRL_ROUTINE_CURRENT();
        }
}
/* USER CODE END 4 */
```

Then, we can build the program and run the program to upload the code to the STM32 microcontroller. Just keep in mind that the system is not modifying the duty cycle directly as a result of the sensed output voltage, it takes and generates a reference current value from the output voltage and the reference current value is what used to actually modify the duty cycle directly, so it is almost indirect control instead of direct control. As you can see on the oscilloscope, the output voltage is not constant and it is going up and down from 11.3 V to 11.5 V, so you need to modify the voltage and current controller coefficients. So, we go back to the code and inside the main.c file we change the set point, V_{ref} value, from 10 V to 15 V. You should note that we can only change the set point V_{ref}

and we cannot change the I_{ref} manually and it is changed automatically and dynamically based on what V_{ref} is, so we run the program gain. There is some offset in the oscilloscope and it is changing from 16.7 V to 16.9 V. So, that is basically what the average current mode control is and it is very useful in a vast majority of applications such as battery chargers and, of course, you cannot use average current mode control if you are not operating in continuous conduction mode, CCM. If the inductors are not large enough, they will not be operating in CCM most time. We should note that we are getting close to the rated, that is, maximum reference current, Iref_max, which is 2.133 A in this case. The last setpoint, V_{ref}, that we want to use is 16 V, so we go to main.c file and set it as given in the code below.

```
/* USER CODE BEGIN PV */
float Vref = 16.0;
float Iref = 0.2;
```

Also, we modify slightly Iref_max from 2.133 to 2.2 in custom.c file as you can see in the following code.

```
//#define      y_v_max         0.75
#define        y_v_max         1.0
#define        y_v_min         0.0
#define        Iref_max        2.2
```

Then, we run the program again, and we see on the oscilloscope that the output voltage goes up and down from 17.9 V to 18.1 V. That is the implementation of average current mode control. The noise in the output voltage is mostly because of the AC/DC converter, using the output of linear voltage regulators as an input to the buck converter will give better results. It is very advantageous to control the current through the inductor and average current mode control can do that very well. If we modify the coefficients, we can reduce the output voltage oscillations, for example, we change the B0_v and B1_v and also B0_i and B1_i values to one-tenth of their previous values in custom.c file as given in the code below.

```
#define         A1_v            1
#define         A2_v            0
#define         A3_v            0
#define         B0_v            0.0001107
#define         B1_v            0.0002056
#define         B2_v            0
#define         B3_v            0
#define         A1_i            1
#define         A2_i            0
#define         A3_i            0
#define         B0_i            0.0008334
#define         B1_i            0.001548
#define         B2_i            0
#define         B3_i            0
```

By doing that, you actually increase the settling time when coefficients are more than what actually they are, and when you reduce them, you actually reduce the settling time and you will have more stable output. Then we run the program, we can see that the output voltage is now more stable and it is not going up and down so much. We again change more the B0_i and B1_i values to one-tenth of their previous values as you can see in the following code.

```
#define         A1_v            1
#define         A2_v            0
#define         A3_v            0
#define         B0_v            0.0001107
#define         B1_v            0.0002056
#define         B2_v            0
#define         B3_v            0
#define         A1_i            1
#define         A2_i            0
```

```
#define        A3_i            0
#define        B0_i            0.00008334
#define        B1_i            0.0001548
#define        B2_i            0
#define        B3_i            0
```

Then, we run the program and can see that the output voltage is now more stable and it is fixed at 18.1 V.

CHAPTER 4

Digital Control Implementation with PLECS

4.1 INTRODUCTION

PLECS is a simulation software tool used for the design and analysis of power electronic systems. It stands for piecewise linear electrical circuit simulation and is developed by Plexim GmbH. PLECS provides a user-friendly interface for modeling complex power electronic systems and simulating their behavior in real-time. Key features of PLECS include:

- Intuitive graphical modeling environment for designing power electronic circuits.
- Support for modeling various power converters, inverters, motor drives, and other power electronic systems.
- Real-time simulation capabilities for rapid prototyping and testing of control algorithms.

DOI: 10.1201/9781003541356-4

- Integration with MATLAB/Simulink for system-level simulation and control design.
- Code generation capabilities for automatically generating embedded code for microcontrollers.
- Extensive library of pre-built power electronic components and control blocks.
- Validation and verification tools for analyzing system behavior and performance.

Overall, PLECS is a powerful tool for engineers and researchers working in the field of power electronics, enabling them to model, simulate, and optimize the performance of power electronic systems efficiently. You can generate code for power converter control in PLECS for STM32 microcontrollers and implement your control algorithm on the microcontroller to regulate the power converter operation. In this chapter, we will investigate and study a practical example of implementing a power converter, designing a controller, and generating code for the STM32 Nucleo-G474RE microcontroller in detail using PLECS and MATLAB software.

4.2 DESIGNING A CLOSED-LOOP CONTROLLER IN PLECS AND MATLAB

In this section, we intend to design a buck converter in the PLECS software environment and simulate it. The designed buck converter is shown in Figure 4.1.

The circuit parameter values are clearly visible in Figure 4.1. The switching frequency is 20 kHz and the triangular wave generator amplitude changes between 0 and 1, and it is compared with constant duty cycle value with a logical comparator or relational operator to generate the PWM signal and, finally, it is fed to

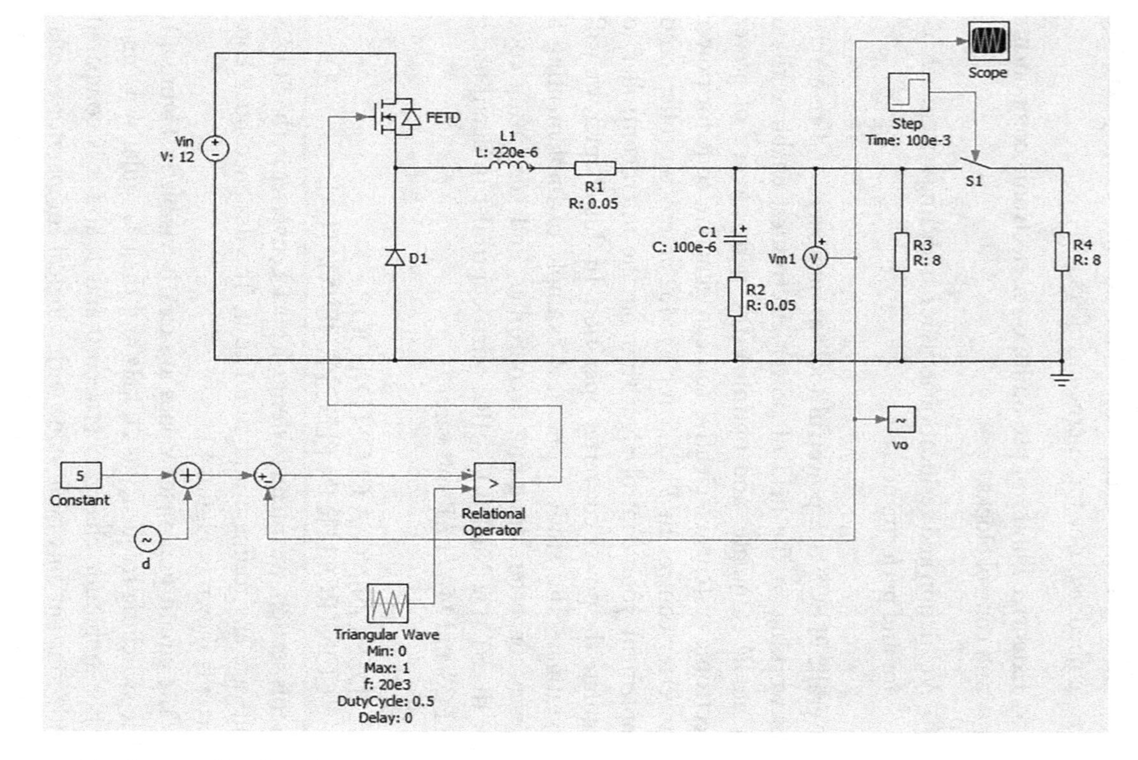

FIGURE 4.1 The designed buck converter in PLECS.

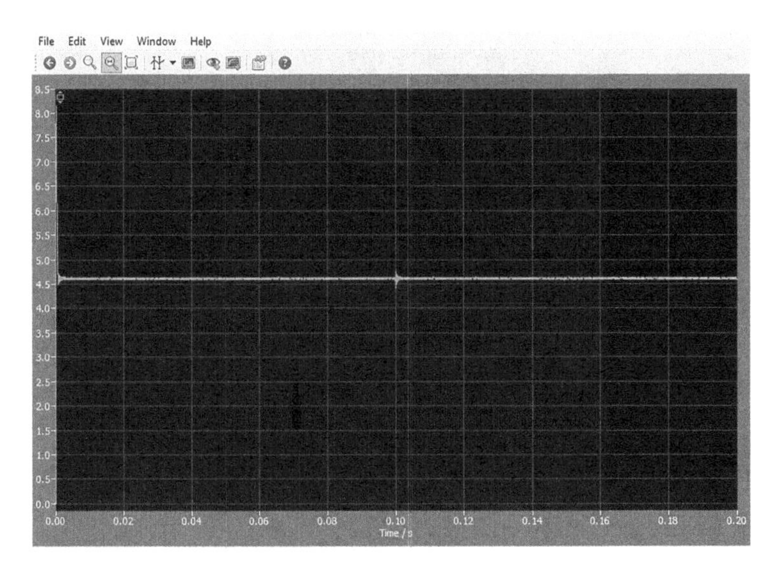

FIGURE 4.2 The output voltage waveform.

the gate driver of the MOSFET. Also, there is a load change at 100 ms and the load resistance is changing from 8 Ω to 4 Ω at the time of 100 ms. The output voltage waveform is also depicted in Figure 4.2 which has an undesirable overshoot at the startup and there is an offset or a deviation from the reference voltage value, 5 V, as shown in Figure 4.2. The simulation parameters window for this system is illustrated in Figure 4.3.

For designing a PID controller for this power converter, we need to know the control to output frequency response of the system. Therefore, in library browser, control, and small signal analysis section, we do drag and drop the small signal perturbation and add it with the reference voltage value and also do drag and drop the small signal response as shown in Figure 4.4 and connect it to the measured output voltage as shown in Figure 4.1.

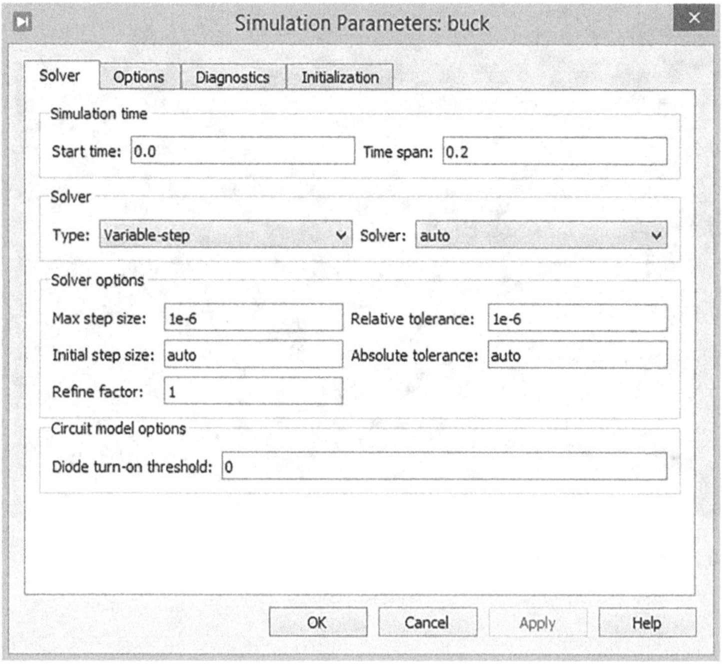

FIGURE 4.3 The simulation parameters window.

To see the control to output frequency response of the system from Simulation tab in menu, we select the Analysis too as shown in Figure 4.5.

In the opened Analysis Tools window, we do click on the plus (+) button and select the AC Sweep Analyses and do the settings as shown in Figure 4.6.

To see the control to output frequency response, that is, Bode diagram of magnitude and phase of the system we do click on the Start analysis button in Analysis Tools window which is depicted in Figure 4.7.

We will use MATLAB software to design a PID controller for closed-loop system so we need to export the frequency response data as a CSV file as illustrated in Figure 4.8.

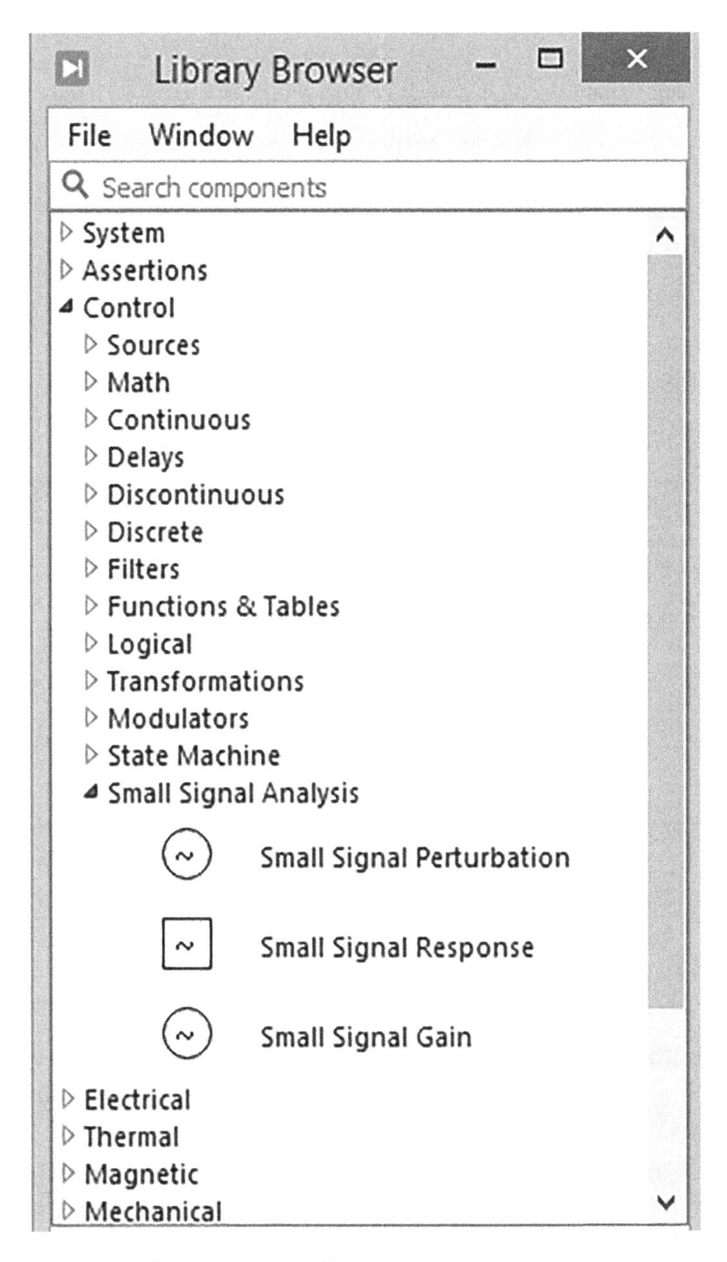

FIGURE 4.4 Choosing the small signal analysis elements.

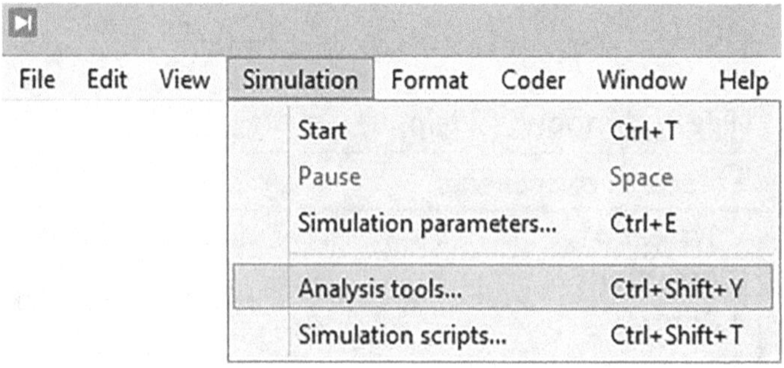

FIGURE 4.5 Selecting the analysis tools… from simulation tab.

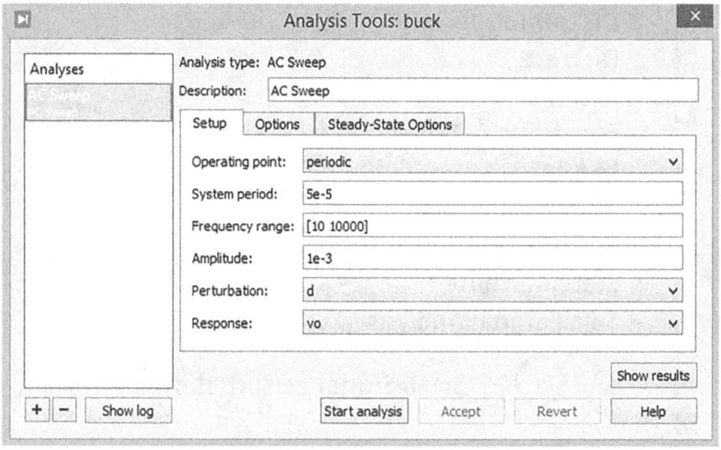

FIGURE 4.6 Settings of the analysis tools window.

In the opened Export as window, we choose a name (data.csv) for it and click on the Save button to save it in the desired directory as shown in Figure 4.9.

Then we open the saved CSV file and delete the header or first row of it and save it again as depicted in Figure 4.10, since if we do not do that when we read the CSV file in MATLAB command window, we will get an error message.

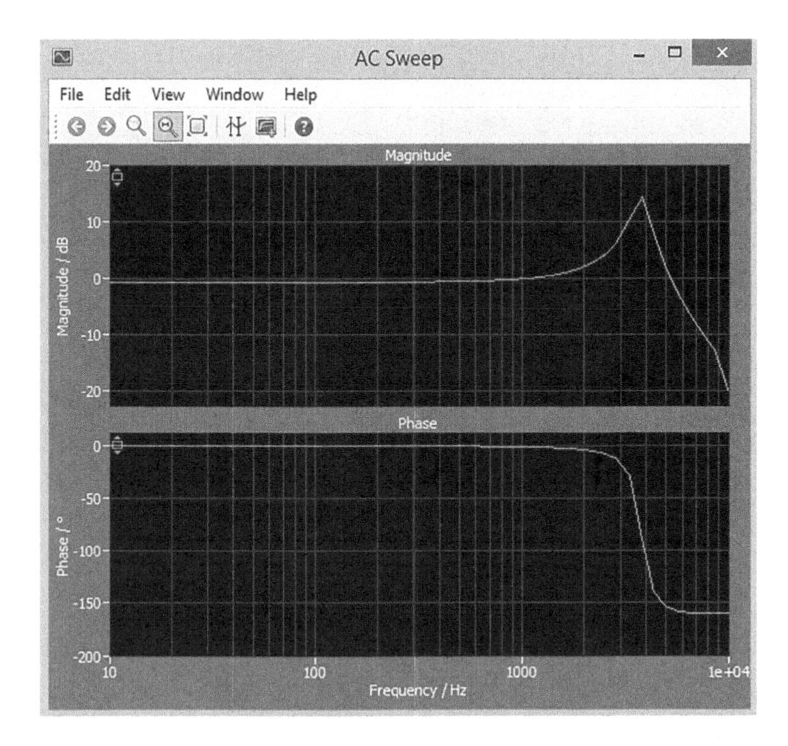

FIGURE 4.7 The control to output frequency response.

Therefore, we use the following codes in the MATLAB command window to read the saved data.csv file and plot the Bode diagram of the system in MATLAB.

```
>> data = csvread('C:\plecs\stm32\data.csv');
>> w = 2*pi*data(:,1);
>> val=10.^(data(:,2)/20).*exp(j*data(:,3)
*pi/180);
>> sys = frd(val,w);
>> bode(sys),grid minor
```

By running those codes in the MATLAB command window, the Bode diagram of the system is plotted in MATLAB as depicted in Figure 4.11.

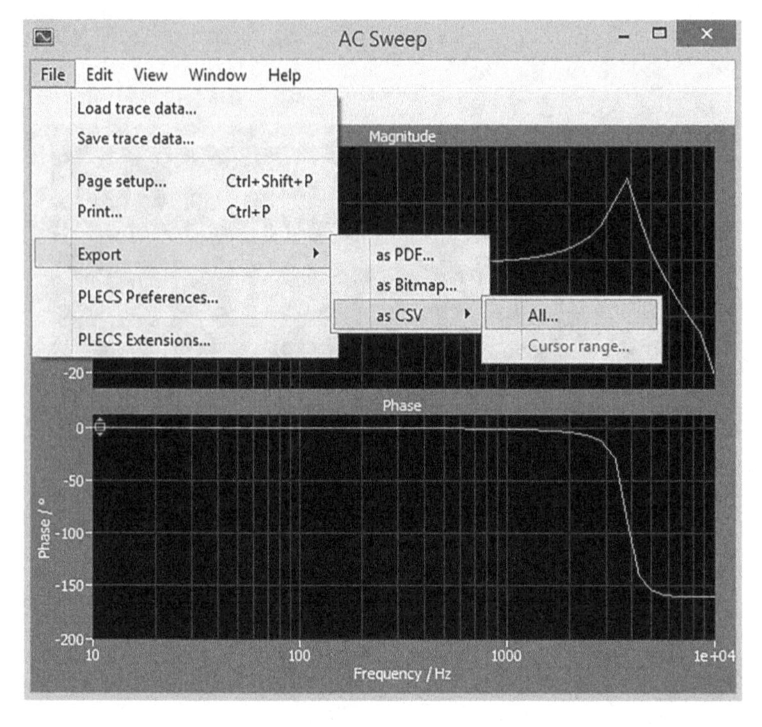

FIGURE 4.8 Exporting the frequency response data as a CSV file.

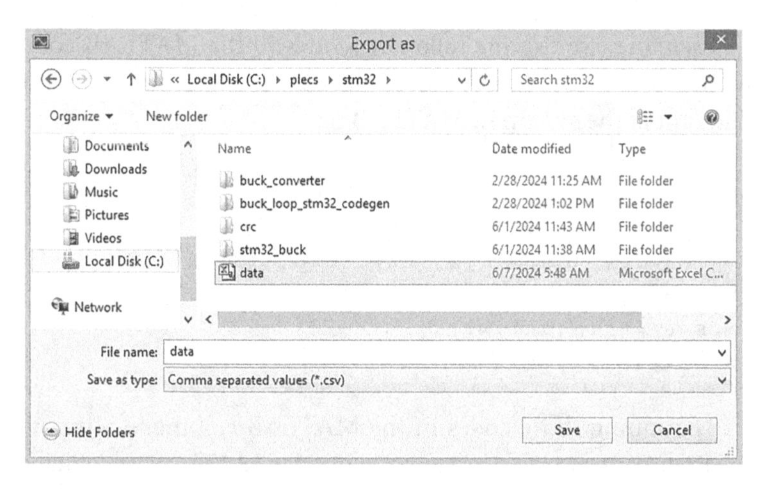

FIGURE 4.9 Choosing a name and saving the CSV file.

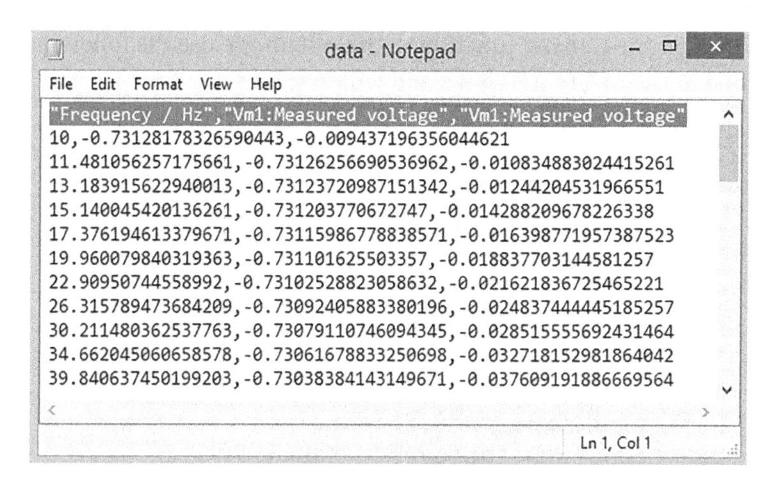

data - Notepad

File Edit Format View Help

```
"Frequency / Hz","Vm1:Measured voltage","Vm1:Measured voltage"
10,-0.7312817832659043,-0.009437196356044621
11.481056257175661,-0.73126256690536962,-0.010834883024415261
13.183915622940013,-0.73123720987151342,-0.01244204531966551
15.140045420136261,-0.731203770672747,-0.014288209678226338
17.376194613379671,-0.73115986778838571,-0.016398771957387523
19.960079840319363,-0.731101625503357,-0.018837703144581257
22.90950744558992,-0.73102528823058632,-0.021621836725465221
26.315789473684209,-0.73092405883380196,-0.024837444445185257
30.211480362537763,-0.73079110746094345,-0.028515555692431464
34.662045060658578,-0.73061678833250698,-0.032718152981864042
39.840637450199203,-0.73038384143149671,-0.037609191886669564
```

Ln 1, Col 1

FIGURE 4.10 Removing the first row of data.csv.

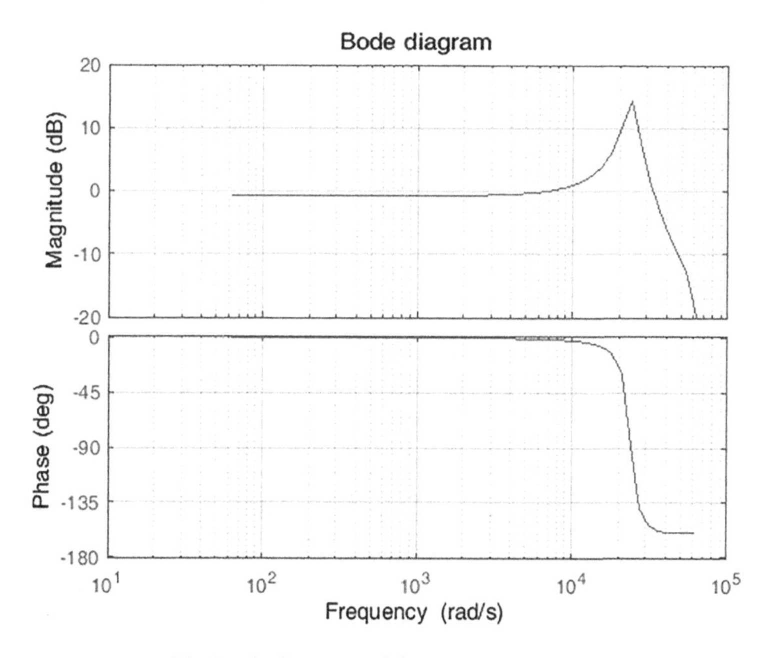

FIGURE 4.11 The Bode diagram of the system in MATLAB.

To get the transfer function of the system, we use the following code in the MATLAB command window.

```
>> H = tfest(sys,2)
H =
   2497 s + 5.186e08
  -----------------------
  s^2 + 4201 s + 5.639e08
 Continuous-time identified transfer function.
Parameterization:
  Number of poles: 2   Number of zeros: 1
  Number of free coefficients: 4
  Use "tfdata", "getpvec", "getcov" for
  parameters and their uncertainties.
Status:
Estimated using TFEST on frequency response data
"sys".
Fit to estimation data: 99.2% (simulation focus)
FPE: 7.625e-05, MSE: 7.07e-05
```

As you can see above, estimated transfer function of the system with an accuracy of 99.2% is as below:

$$H = \frac{2497s + 5.186 \times 10^8}{s^2 + 4201s + 5.639 \times 10^8} \tag{4.1}$$

Then we use the following code in MATLAB command window to design a closed-loop PID controller for the system.

```
>> pidTuner(H)
```

By running the code above, the PID Tuner window is opened and we select the Type as PID and Form as Parallel and also set the Response Time (0.1173) and Transient behavior (0.66), so that an appropriate step plot is achieved as shown in Figure 4.12.

Then we click on the Export button in the PID Tuner window and in the opened Export Linear System we do enter a name (PID) in Export PID controller and click on the OK button as illustrated in Figure 4.13.

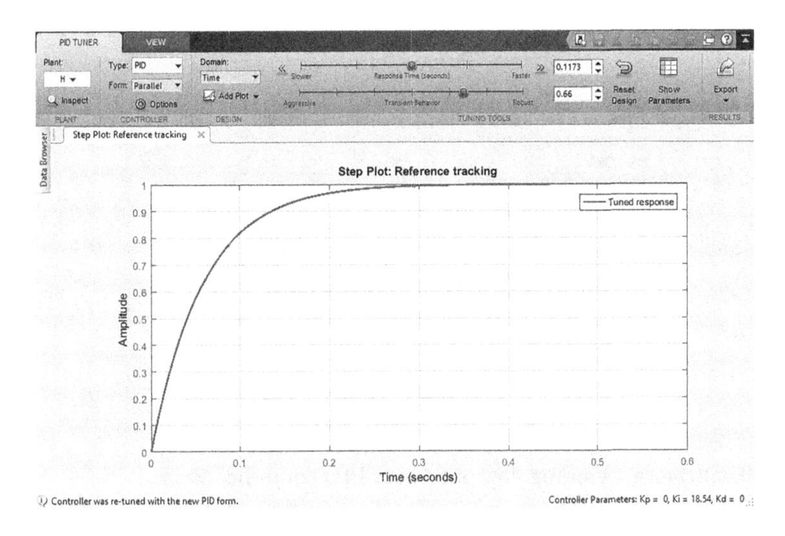

FIGURE 4.12 Setting the PID tuner window.

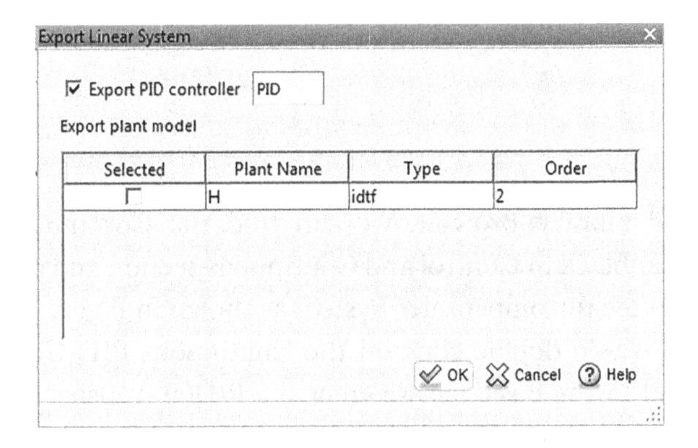

FIGURE 4.13 Entering a name (PID) in export PID controller window.

Finally, in the MATLAB command window, we run the code below to get the closed-loop PID controller transfer function.

```
>> PID
PID =
      1
  Ki * ---
      s
  with Ki = 18.5
 Continuous-time I-only controller.
```

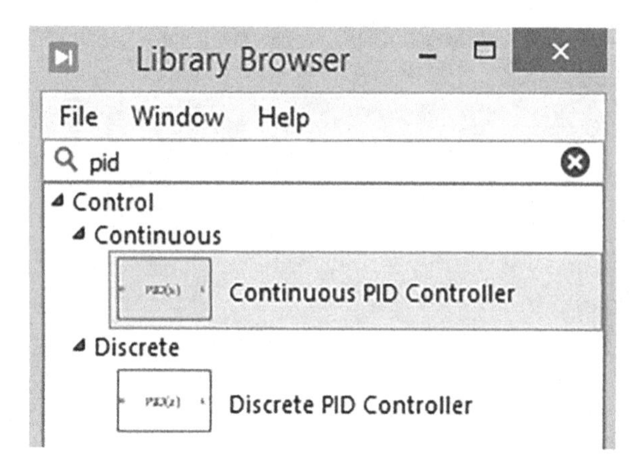

FIGURE 4.14 Finding the continuous PID controller block.

So, the designed closed-loop PID controller transfer function is as follows:

$$\mathrm{PID}(s) = \frac{18.5}{s} \tag{4.2}$$

In the Library Browser, we can find the Continuous PID Controller block in Control and Continuous section and drag and drop it to the uncompensated system as shown in Figure 4.14.

Then, we do double click on the Continuous PID Controller block and in the Basic tab we enter the PID(s) transfer function values as depicted in Figure 4.15.

Also, in the Anti-Windup tab of the Continuous PID Controller window, we set the upper saturation limit to 0.9 and lower saturation limit to 0.1, as shown in Figure 4.16, to limit the duty cycle between 0.1 and 0.9 in order to safe operation of the MOSFET. Then we can add the Continuous PID Controller to the closed-loop control system as illustrated in Figure 4.17.

Finally, we simulate the compensated closed-loop control system and we can see the output voltage waveform on the Scope as shown in Figure 4.18.

As you can see in this case the output voltage offset is omitted and the output voltage is equal to the reference voltage value

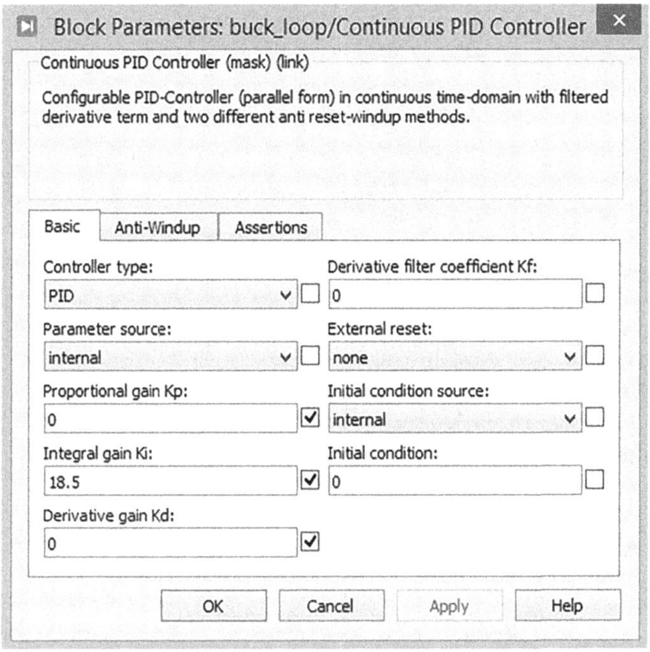

FIGURE 4.15 Setting the PID(s) transfer function values.

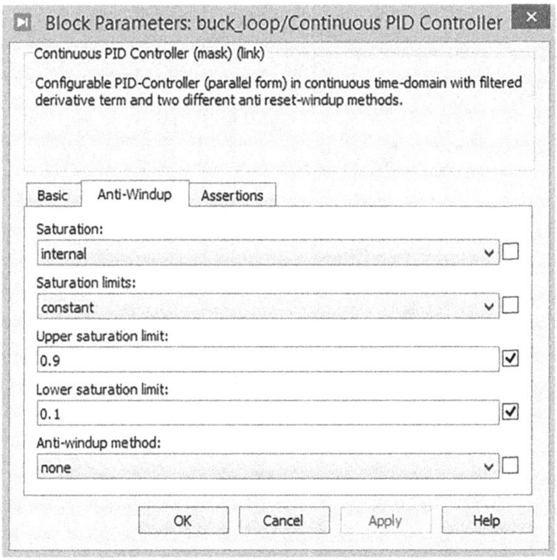

FIGURE 4.16 Setting the upper and lower saturation limits in the anti-windup tab.

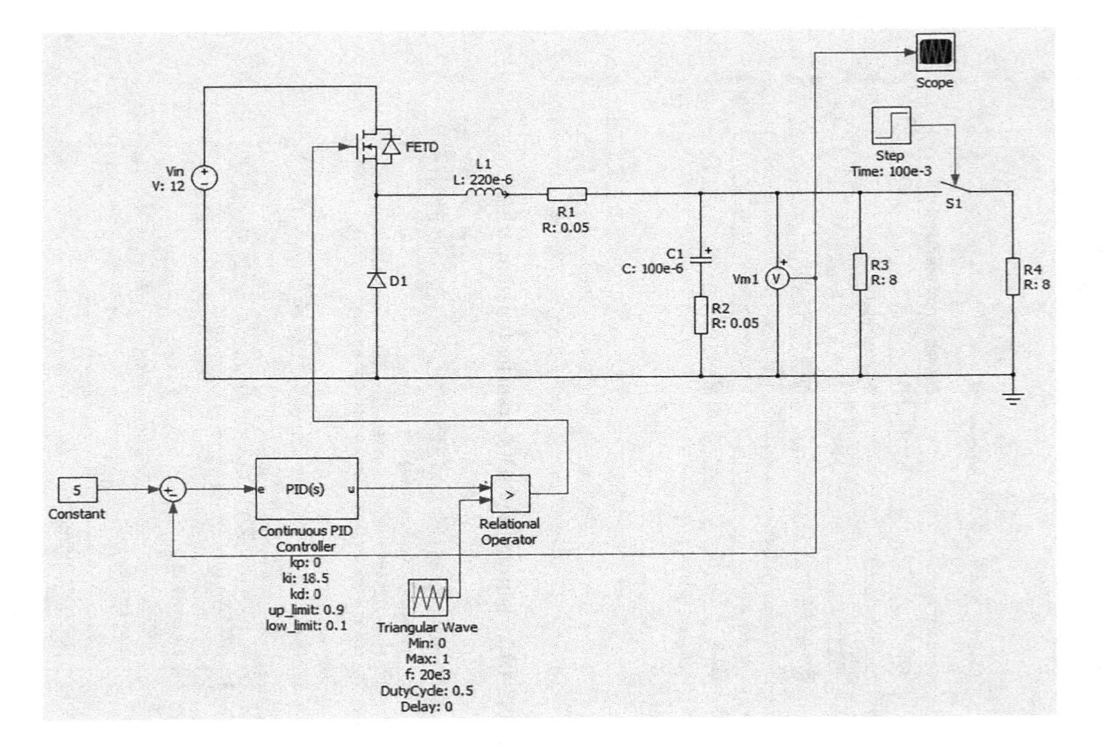

FIGURE 4.17 Adding the PID(s) block to the closed-loop control system.

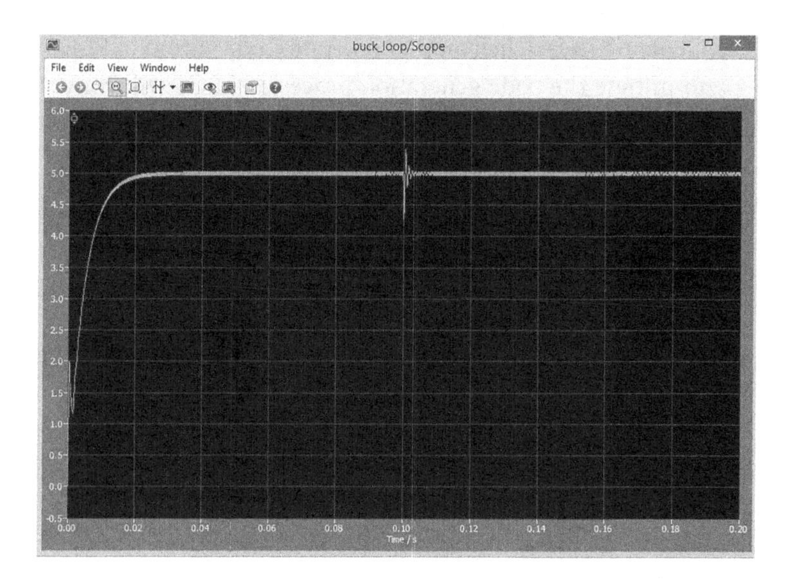

FIGURE 4.18 The output voltage waveform on the scope.

which is 5 V. Also, the startup overshoot in output voltage is removed and following a step load resistance change from 8 Ω to 4 Ω at 100 ms, the output voltage becomes stable in less than 5 ms.

4.3 CODE GENERATION FOR THE STM32 NUCLEO-G474RE IN PLECS

Code generation for the STM32 Nucleo-G474RE in PLECS involves converting the simulation model created in PLECS into executable code that can be run on the specific microcontroller. Here is a general overview of the steps involved in code generation for the STM32 Nucleo-G474RE in PLECS:

1. Design the control system in PLECS: Create and simulate the control system model in PLECS to ensure that it meets the desired performance specifications.

2. Configure the target settings: Select the STM32 Nucleo-G474RE microcontroller as the target for code generation in PLECS by specifying the appropriate settings in the Target Setup window.

3. Generate code: Click on the Generate Code button in PLECS to initiate the code generation process. PLECS will convert the simulation model into C code that can be compiled and uploaded to the STM32 Nucleo-G474RE.

4. Compile and upload code: Compile the generated C code using an Integrated Development Environment (IDE) such as the STM32CubeIDE. Upload the compiled code to the STM32 Nucleo-G474RE microcontroller using a programming tool such as ST-Link.

5. Test the control system: Run the control system on the STM32 Nucleo-G474RE to verify its functionality and performance. Make any necessary adjustments to the code or simulation model in PLECS to optimize the system's performance.

By following these steps, you can effectively generate code for the STM32 Nucleo-G474RE microcontroller in PLECS and implement the control system on the microcontroller for real-world applications. Now, we are going to create the same buck converter using the STM32 PLECS library and then generate the code for the STM32 Nucleo-G474RE microcontroller in STM32CubeIDE. To do that, we create two subsystems one for the buck converter and the other one for the PID controller as shown in Figure 4.19.

The circuit inside the Buck_Converter subsystem is depicted in Figure 4.20.

As you can see in Figure 4.20, we have added a resistor divider with the values R5 = 100 kΩ and R6 = 10 kΩ, since we want to get

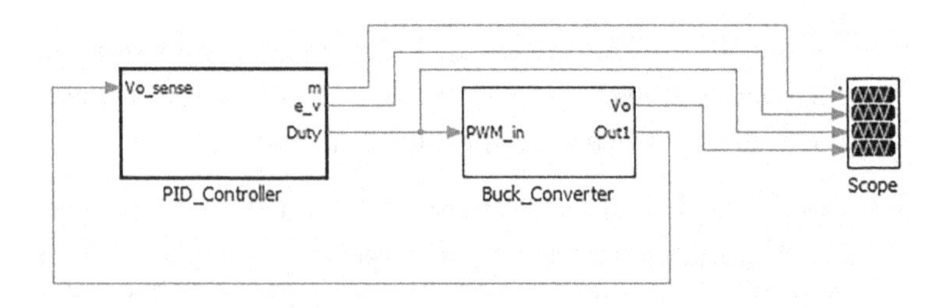

FIGURE 4.19 Creating two subsystems for buck converter and PID controller.

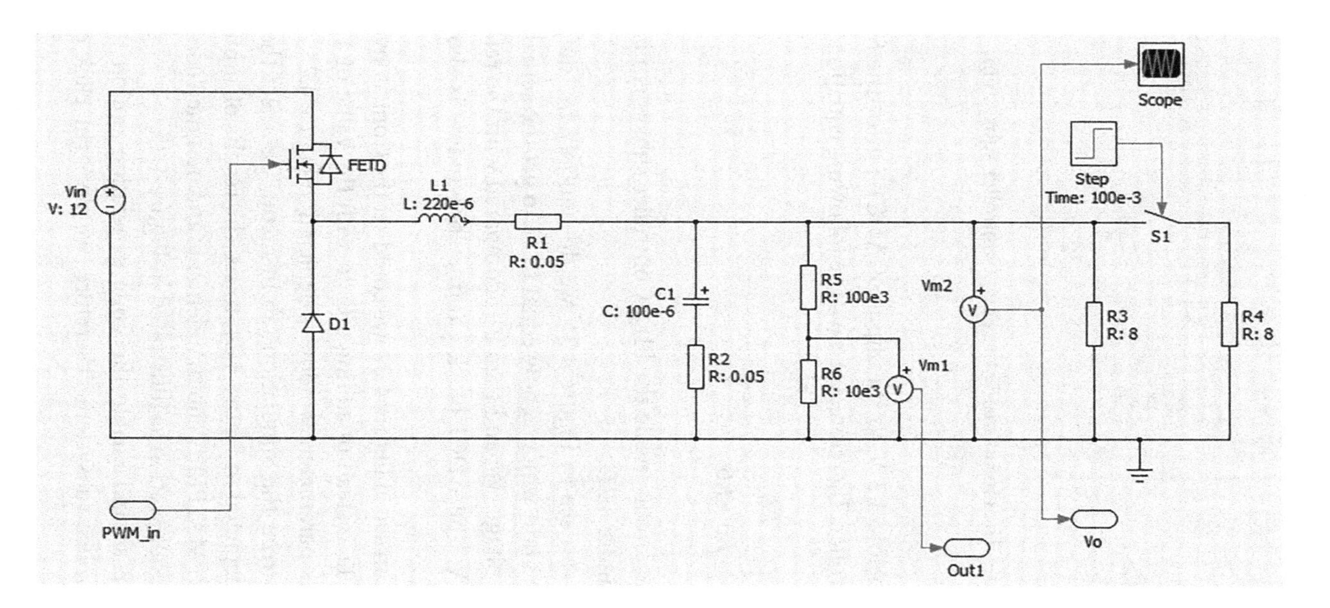

FIGURE 4.20 The circuit inside the Buck_Converter subsystem.

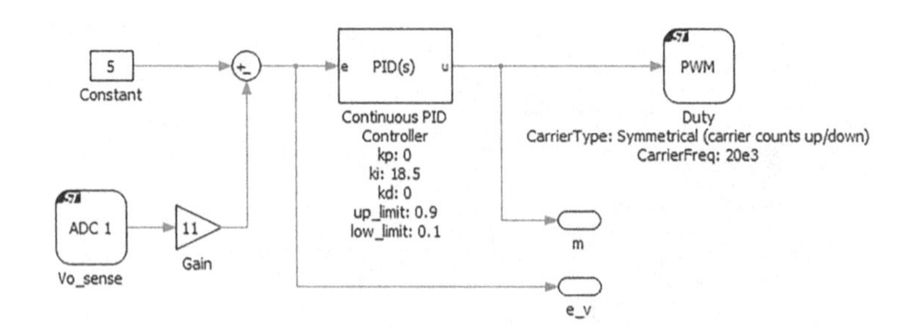

FIGURE 4.21 The circuit inside the PID_Controller subsystem.

a voltage below 3.3 V to the input of ADC1 pin of the STM32 microcontroller. Therefore, we have the following equation:

$$Vo_{\text{sense}} = \frac{R6}{R5 + R6} \times Vo = \frac{1}{11} Vo \tag{4.3}$$

Also, the circuit inside the PID_Controller subsystem is illustrated in Figure 4.21.

As you can see in Figure 4.21, we do multiply the output of ADC1 by a Gain which is now equal to 11 to get the actual value of output voltage, *Vo*, and then it is compared with the reference voltage, 5 V. The ADC1 block parameters window is shown in Figure 4.22.

As you can see in Figure 4.22, we could set the Cont. conversion scale(s) to 11 instead of adding a Gain with the value of 11. The PWM block parameters window is depicted in Figure 4.23.

Finally, from the Simulation tab in menu, we select the Start option as depicted in Figure 4.24, to see the results of simulation using the STM32 PLECS library elements and designed controller in the previous section as illustrated in Figure 4.25.

Then we should enable the code generation option for the PID_Controller subsystem. Therefore, we do right click on the

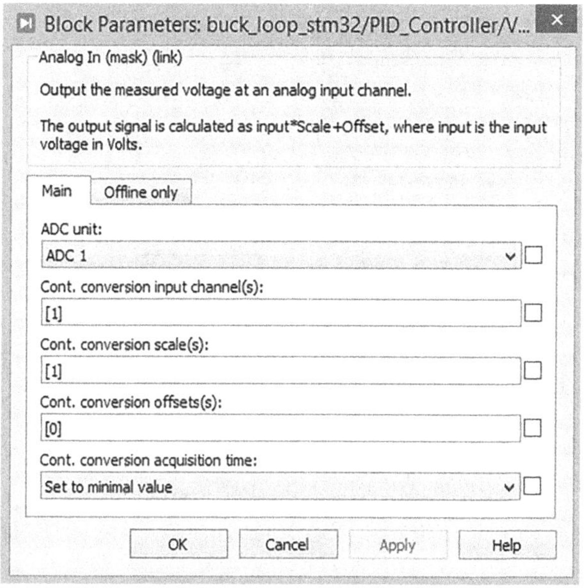

FIGURE 4.22 The ADC1 block parameters window.

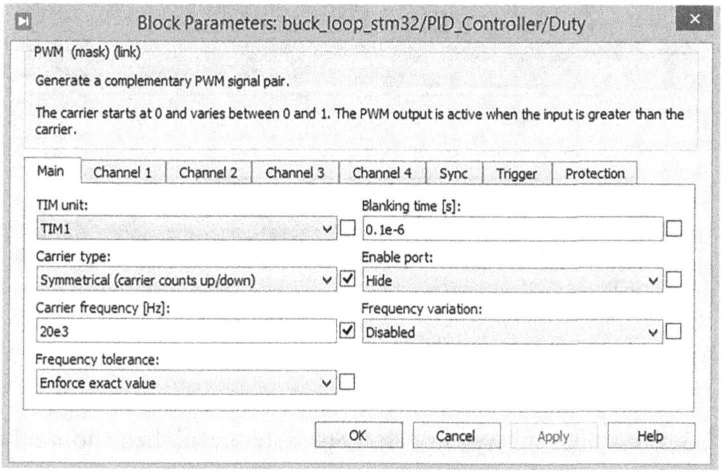

FIGURE 4.23 The PWM block parameters window.

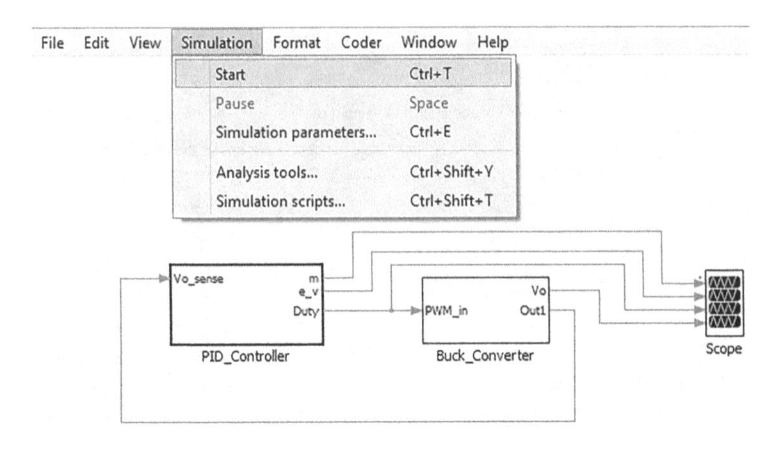

FIGURE 4.24 Selecting the start option.

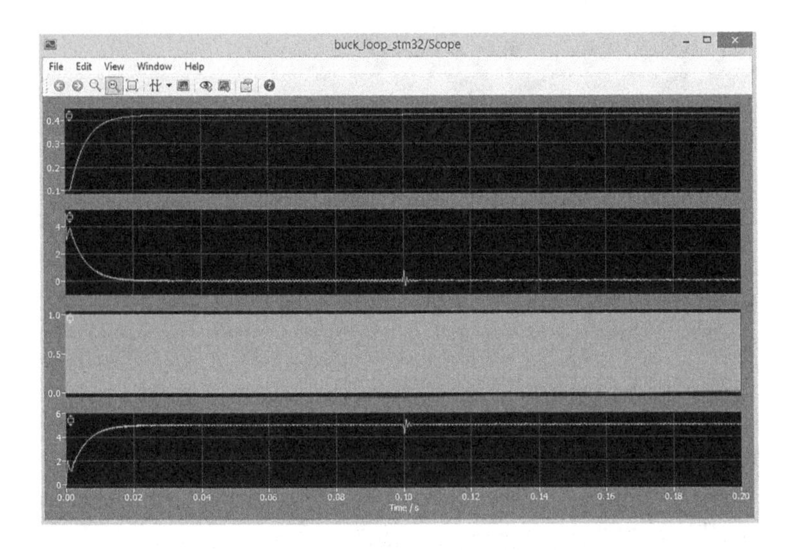

FIGURE 4.25 The results of simulation.

PID_Controller subsystem, select Subsystem, and then choose the Execution settings… option as shown in Figure 4.26.

In the opened Execution Settings window, we tick on the Enable code generation and enter the value of 1e-4 in Discretization step size field and then click on Apply and OK buttons, respectively, as shown in Figure 4.27.

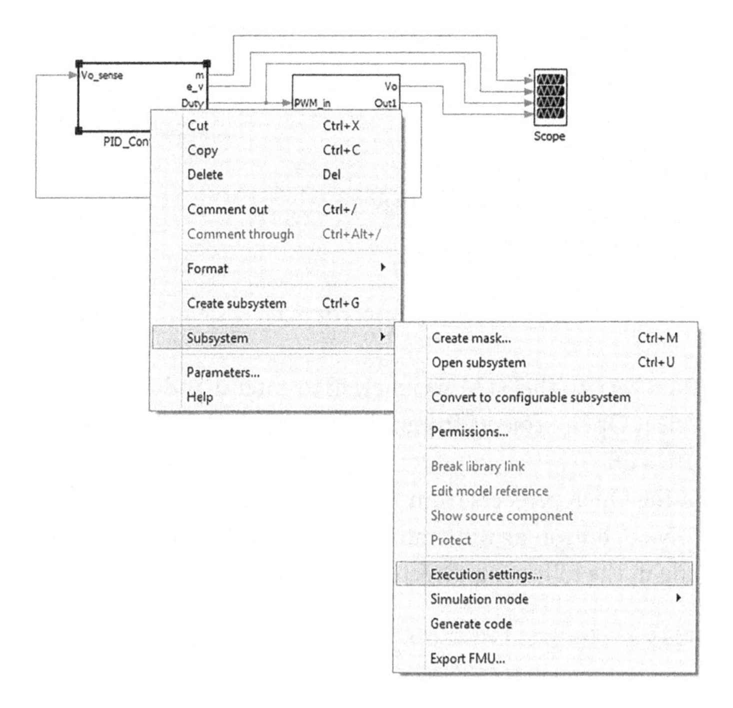

FIGURE 4.26 Choosing the execution settings… option.

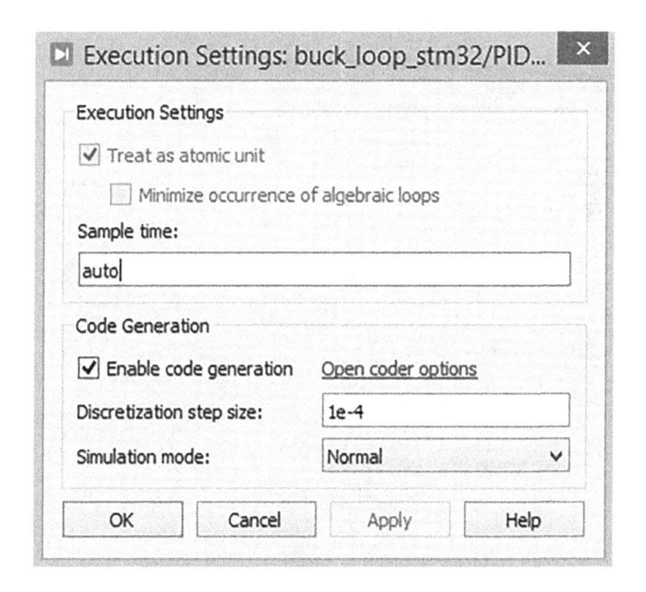

FIGURE 4.27 Ticking on the enable code generation.

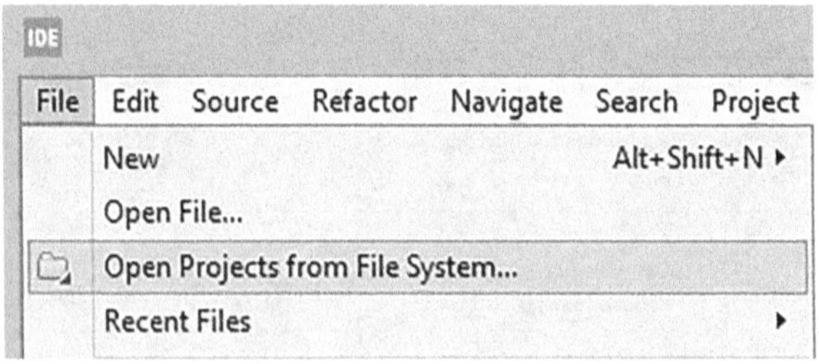

FIGURE 4.28 Selecting open projects from file system… option.

Then we run the STM32CubeIDE to enter it and from File icon we select Open projects from File System… option as depicted in Figure 4.28.

In the Open projects from File System window, we click on the Archive… button as illustrated in Figure 4.29, to open the g474.zip file in the following directory as shown in Figure 4.30.

```
PLECS4.7(64 bit) > tsp_stm32 > projects > g474.zip
```

Then we do click on the Finish button in the Import Projects from File System or Archive window as shown in Figure 4.31.

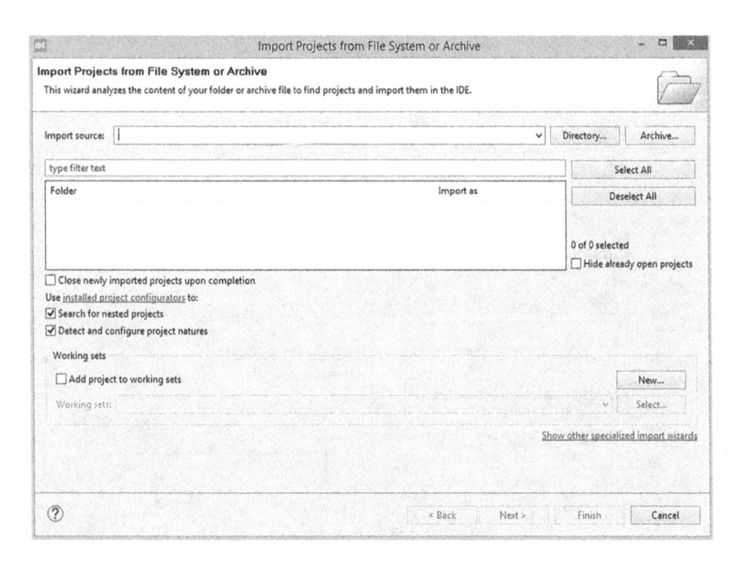

FIGURE 4.29 Clicking on the archive… button.

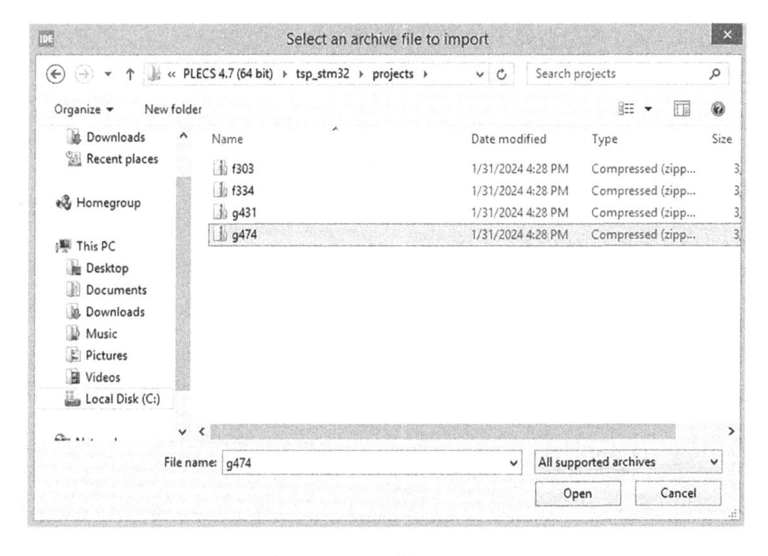

FIGURE 4.30 Opening the g474.zip file.

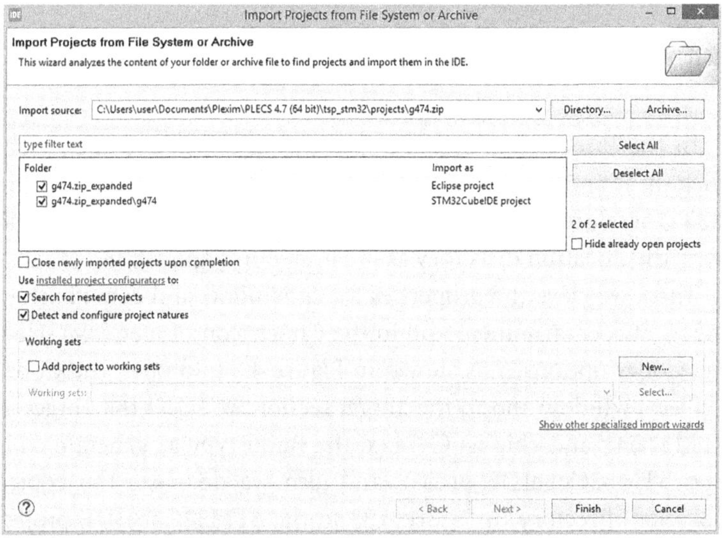

FIGURE 4.31 Clicking on the finish button.

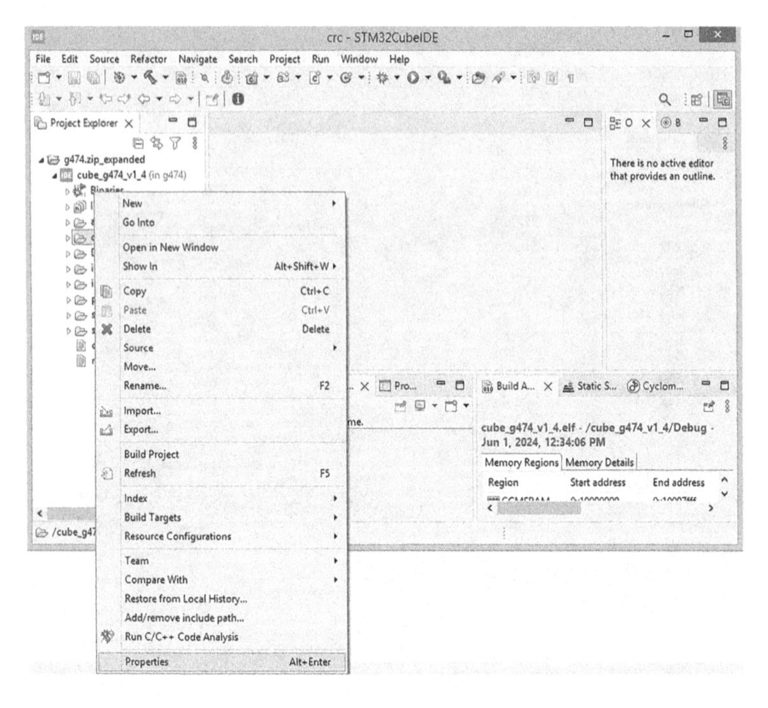

FIGURE 4.32 Selecting the properties option.

In the Project Explorer, we find the cg folder and then right click on it and select the Properties option as depicted in Figure 4.32. Therefore, the Properties for cg window is opened and we do copy the Location directory as illustrated in Figure 4.33.

Then, we close the properties for cg window and come back to the PLECS environment and in the Coder icon of menu we select the Coder options... as shown in Figure 4.34, to open the Coder Options window and in the Target section we select the Target as STM32G4x and chip as G474xx, the Build type as generate code into STM32CubeIDE project and also we do paste the copied Location directory in Figure 4.33 to the STM32CubeIDE project directory and also tick use internal oscillator and finally click on the Accept and Build buttons, respectively, as shown in Figure 4.35.

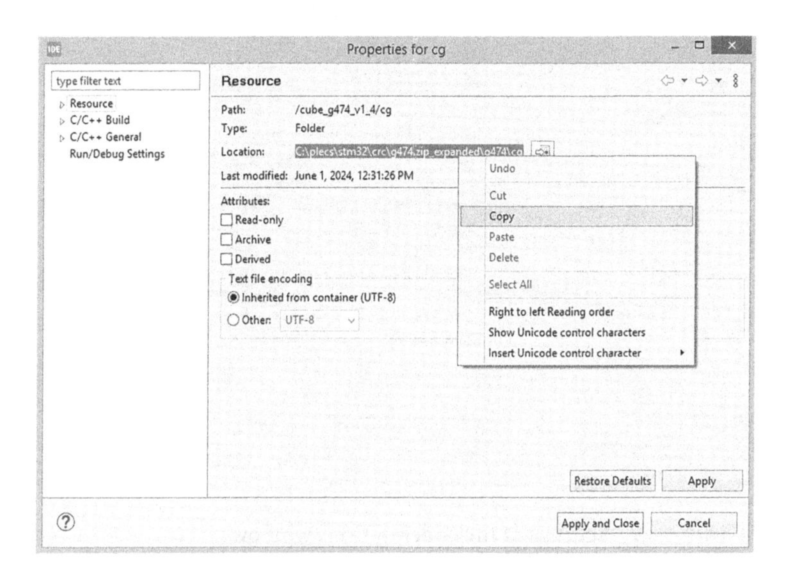

FIGURE 4.33 Doing copy the location directory.

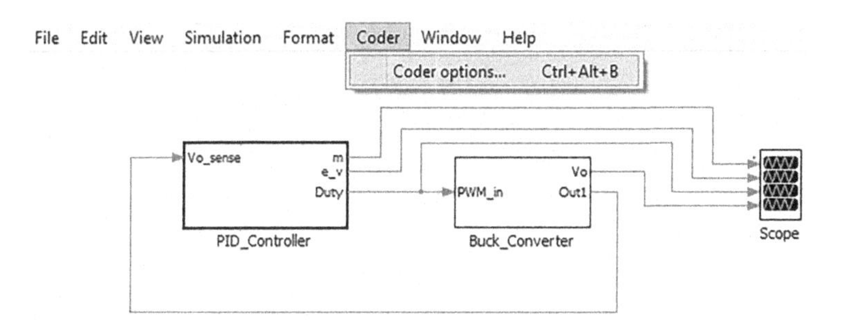

FIGURE 4.34 Selecting the coder options....

As shown in Figure 4.36, we get an error message since we have not defined the I/O components for the output ports, m and e_v, of the PID_Controller subsystem. We should note that we used both of them for simulation purpose and in code generation we can remove them from the PID_Controller subsystem as depicted in Figures 4.37 and 4.38, respectively.

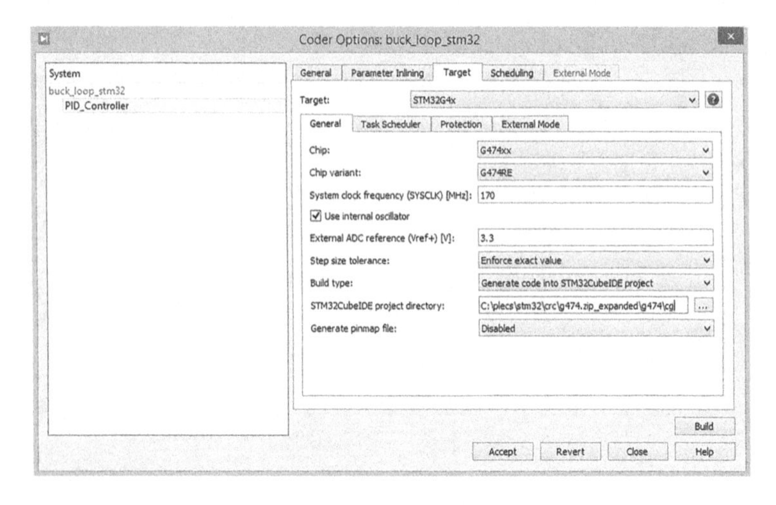

FIGURE 4.35 Settings of the coder options window.

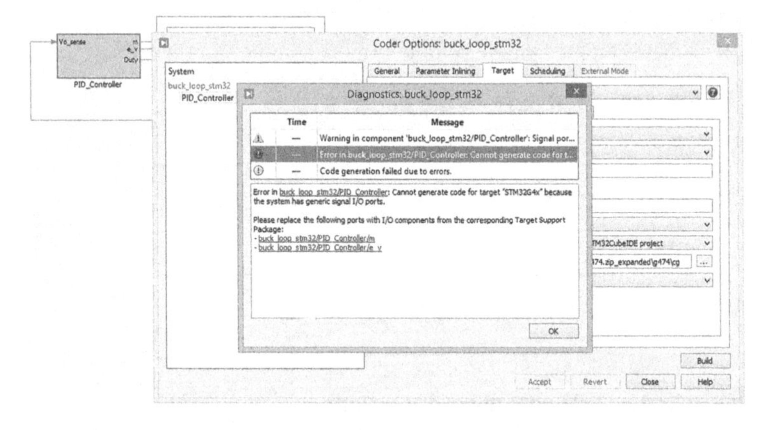

FIGURE 4.36 Getting an error message.

Again, in the Coder Options window, we do click on the build button as illustrated in Figure 4.35. The build is done successfully and we come back to the STM32CubeIDE and in the Project Explorer section we do right click and select the Refresh option as illustrated in Figure 4.39 to see the codes added to the cg folder as shown in Figure 4.40.

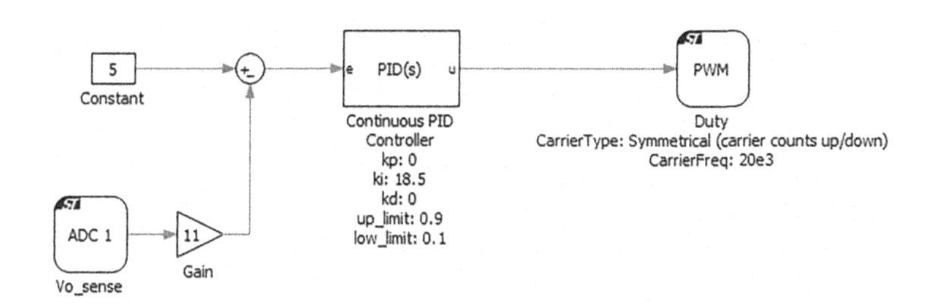

FIGURE 4.37 Removing e_v and m ports from the PID_Controller subsystem.

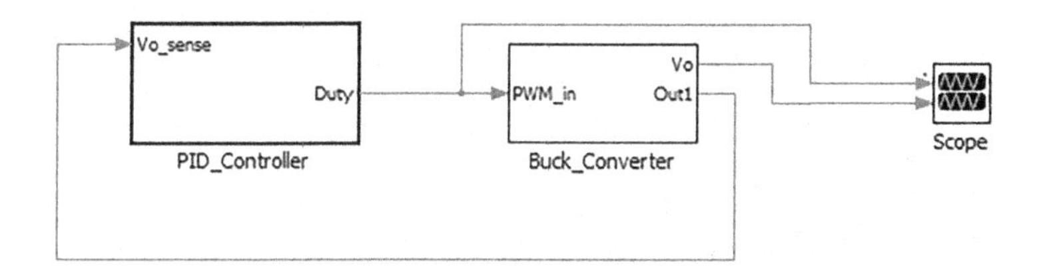

FIGURE 4.38 The PID_Controller subsystem without e_v and m ports.

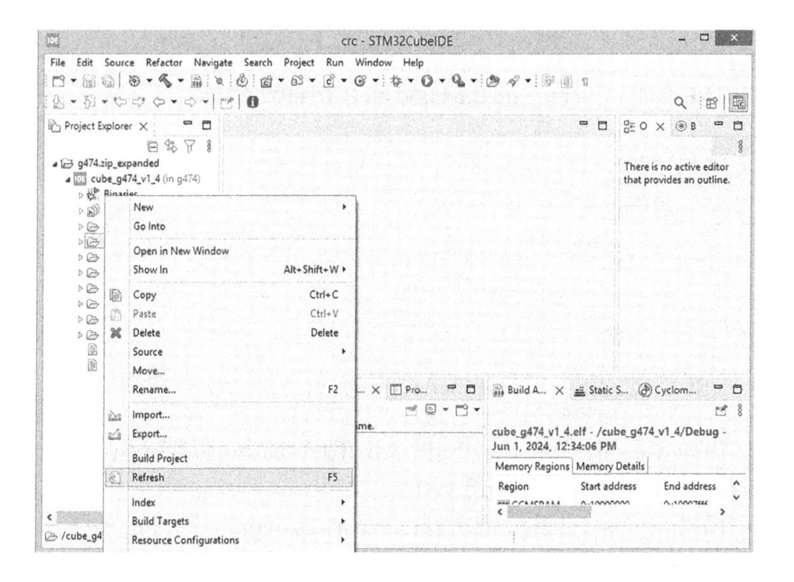

FIGURE 4.39 Selecting the refresh option.

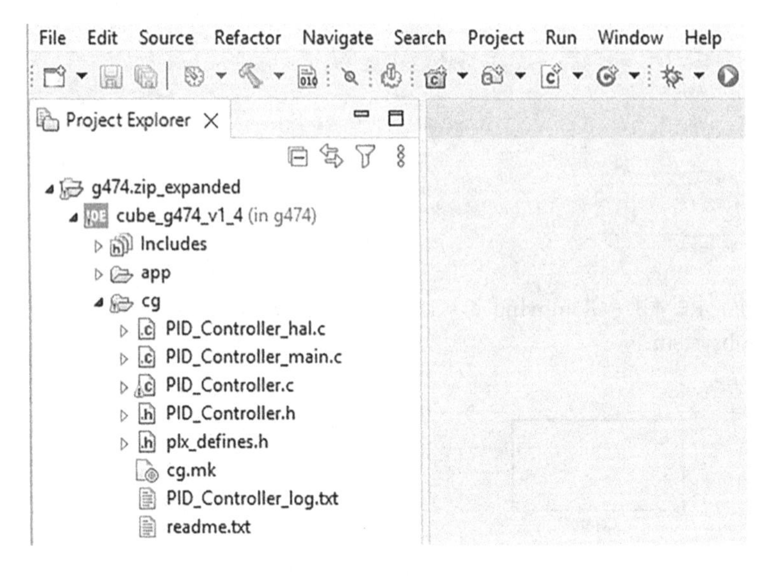

FIGURE 4.40 The codes added to the 'cg' folder.

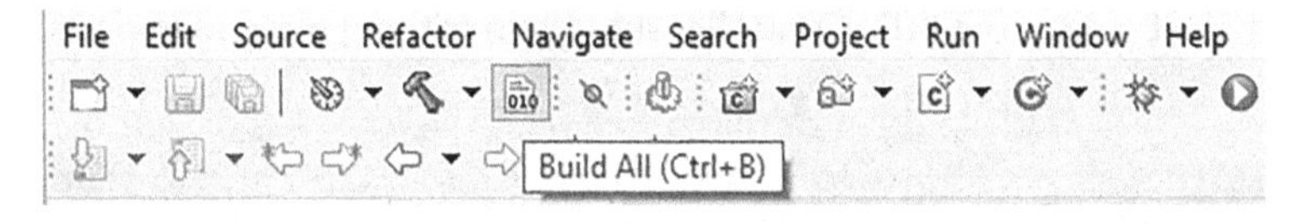

FIGURE 4.41 Clicking on the build all (Ctrl+B) button.

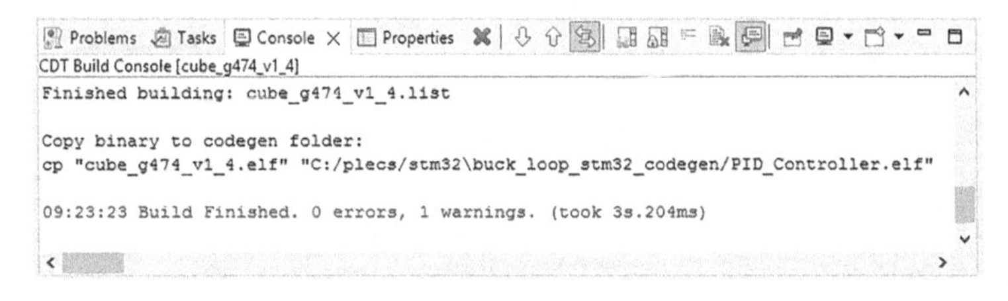

FIGURE 4.42 The build finished message.

Then we click on the Build All (Ctrl+B) button to build the project as shown in Figure 4.41.

The build is finished with zero errors as depicted in Figure 4.42.

Finally, we can run the program to upload the codes to the STM32 microcontroller as illustrated in Figure 4.43.

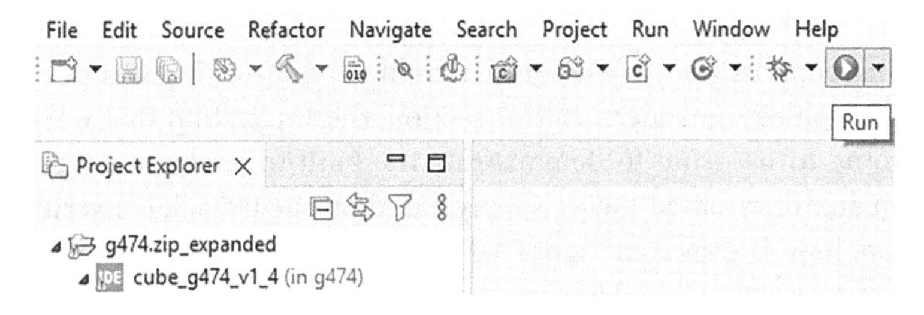

FIGURE 4.43 Running the program.

4.4 REVIEW OF SMALL SIGNAL ANALYSIS TOOLS FOR CONTROLLER DESIGN IN PLECS

We're going to review how you can use the small signal analysis tools found in PLECS for controller design. Since the focus is going to be on using the analysis tools, we may not comment on every single detail within our PLECS circuits that we're going to be using. The electronic engineering software, PLECS, is quite widely used and adopted in various industries and in academic institutions worldwide and it is a complete power conversion system simulation package. It yields very robust and fast simulation results and it's a fully integrated power electronic system simulation package. PLECS standalone offers an independent solution. PLECS features a comprehensive component library, which covers not only the electrical aspects but also the magnetic mechanical and thermal aspects of power conversion systems and their associated controls. So, let's quickly go over the development tools that we'll actually be showing in this section. So, we'll be using PLECS standalone to demonstrate the capabilities of the built-in analysis tools models will contain components from the electrical component library, which includes both passive and active electrical components like resistors and semiconductors. The magnetic, thermal, and mechanical modeling domains are also available in PLECS. With the help of the built-in analysis tools and the controls library, we will design and model our controller. PLECS does offer other tools that further support the simulation development and testing of power electronics controllers. In this section, the application that we're going to be using to demonstrate the built-in analysis tools is an auxiliary-based power supply based on the flyback converter topology as shown in Figure 4.44.

So, because we will be saving some time by using a few prebuilt models, we'll take a moment here to review what's already been developed. As we open new models, we'll explain the changes that we've made. So hopefully it's easy for you to follow the flow. As an input to the converter, here in Figure 4.44, we have a single-phase rectifier. We're using a simplified high-frequency transformer with three output windings, 5 V and plus and minus 12 V. The total output power is around 30 W. This is a very low-cost simple implementation. It's a popular choice for power supply designs. The topology provides isolation between the primary and the secondary sides, and it also then gives the designer the ability to provide multiple outputs for different voltage levels and a choice of positive or negative voltage for the output. So, during this section, we will add an inner peak current controller, which will regulate the peak MOSFET current and then also a voltage controller to regulate the 5-V winding. As we mentioned, we're going to present these control analysis tools within the context of a design example. So, for the current and voltage control loop design of the flyback converter, we can use the procedures depicted in Figure 4.45.

This section will use a design flow that works well within the PLECS modeling environment using these tools. There are of course other approaches and methodologies that are possible. So as a starting point, we will determine the small signal transfer function of the current-controlled converter in the form of generating a Bode plot, and this will then allow us to design a voltage controller and then subsequently calculate the loop gain

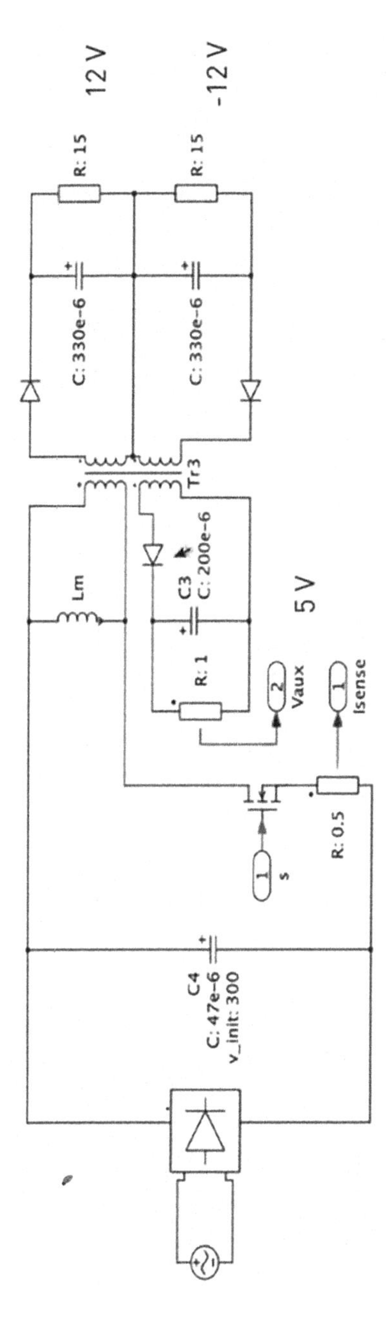

FIGURE 4.44 The topology of auxiliary-based power supply.

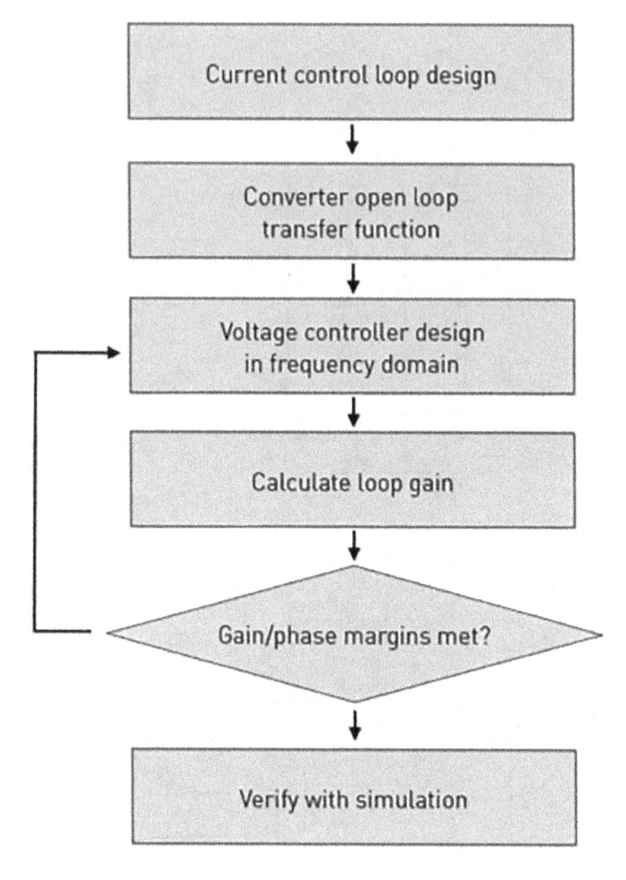

FIGURE 4.45 The current and voltage control loop design of the flyback converter.

response to ascertain if that system is now stable. So as a quick review here, let's go over the peak current controller block, which is actually found in the PLECS library as illustrated in Figure 4.46.

So, this is a modifiable controller. It's part of the controls category. We'll be opening it in PLECS, allowing you to see how it is based on an SR latch or flip-flop storage circuit. It has also a slope

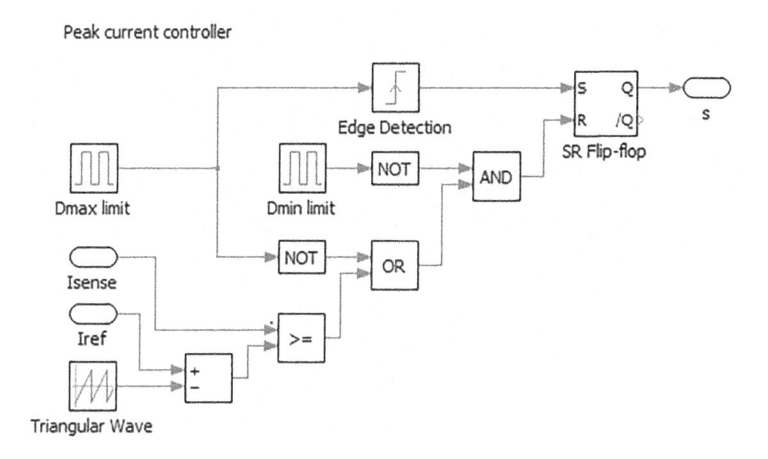

FIGURE 4.46 The peak current controller.

control to avoid oscillations when duty cycle, *d*, is greater than 50% ($d > 50\%$). At the beginning of each switching cycle, a pulse is applied at the input at the set input S of the flip-flop, which also turns the switch on. The switch signal remains on until the maximum duty cycle as specified by the user in the controller's parameters is reached or the current sent signal exceeds the current reference signal. The slope control is also included in the peak current mode controller to avoid oscillations if the duty cycle is greater than 50% and it does this by subtracting a ramp signal from the reference current input. Let's get to work now. We'll open up PLECS as we mentioned standalone and you should now see a model in front of you in the PLECS schematic as shown in Figure 4.47.

So, here in Figure 4.47, we do have a model of that flyback converter with no closed-loop control at all. The gate of our MOSFET is connected to a pulse generator block and we'll first run an open-loop simulation here so you can see the output voltage waveform. We'll do this by going to simulation and then select the start option,

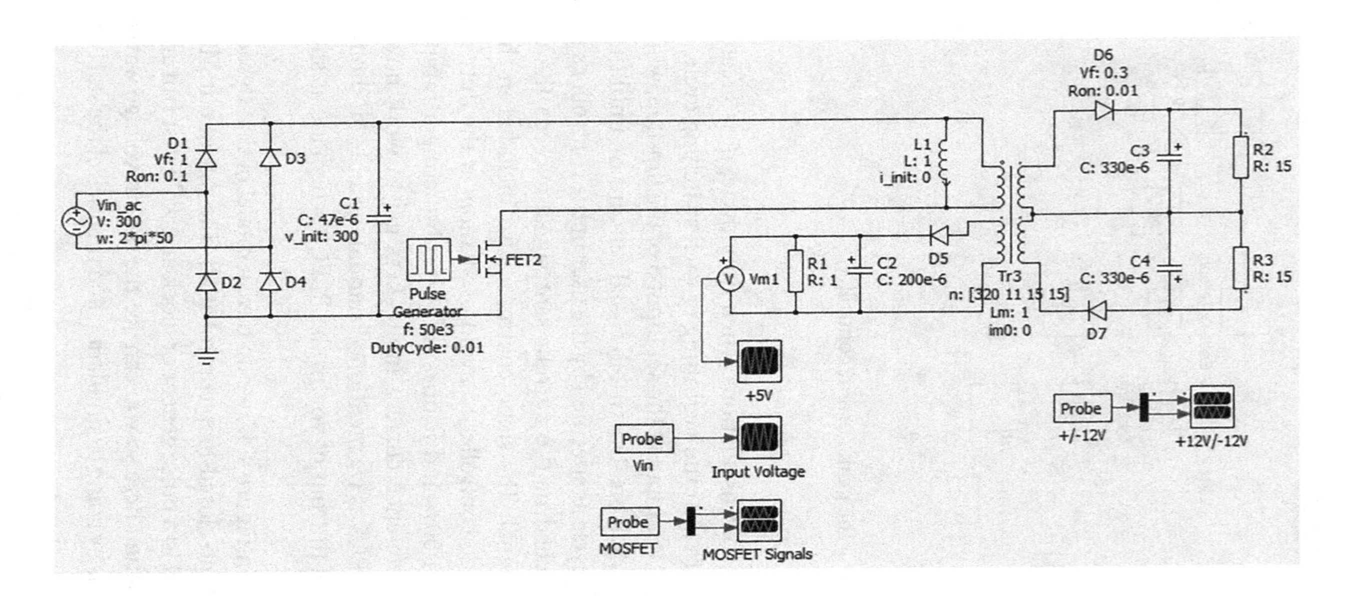

FIGURE 4.47 The flyback converter model in PLECS schematic.

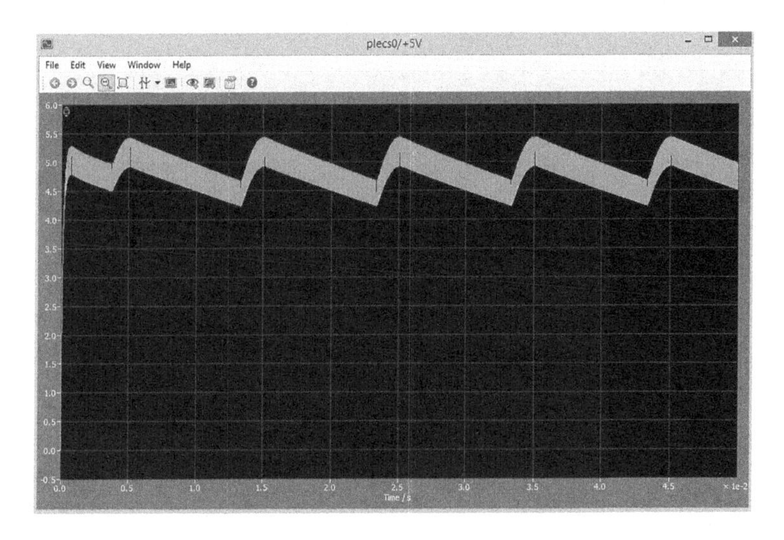

FIGURE 4.48 The voltage waveform of +5 V scope.

and as shown in Figure 4.48 that the startup transients are really not very smooth. Hence, we need to add a voltage controller.

This is what we're going to be working on. In the PLECS library browser, we're going to add components in our library to our schematic and to implement our control architecture. So first, we should replace the pulse generator block with the peak current controller that we've been talking about. So, we can use the search function and search for the component and then we drag it into our schematic. First, we'll connect the switch signal to the MOSFET, and now we can right click this component as with all components in the library that are subsystems. You can select the look under mask option as shown in Figure 4.49, and you can see the implementation as in Figure 4.46, which we've modeled it and you can then modify it as you would for your own application.

So, we can see here the behavior described in Figure 4.46. If we double click the peak current controller, we can see the masked subsystem parameters, so the only default parameters here that we're going to change are just the duty cycle, which we will set to

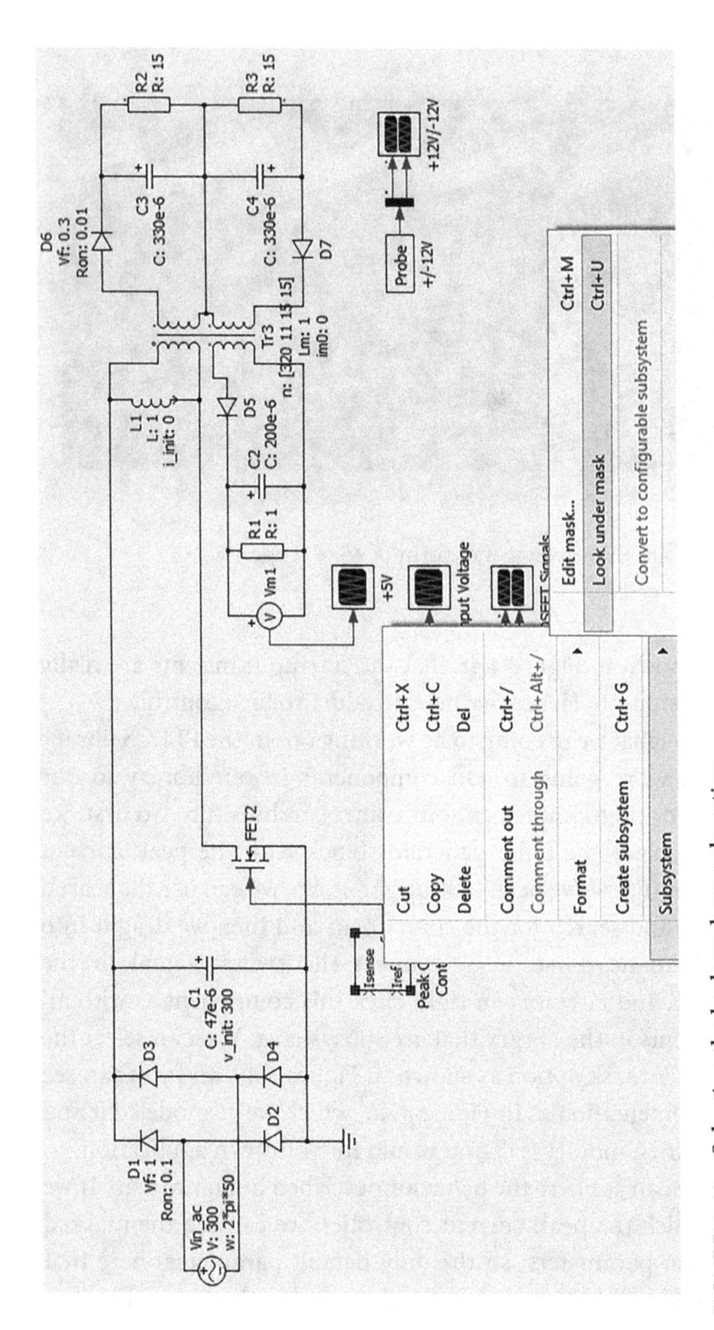

FIGURE 4.49 Selecting the look under mask option.

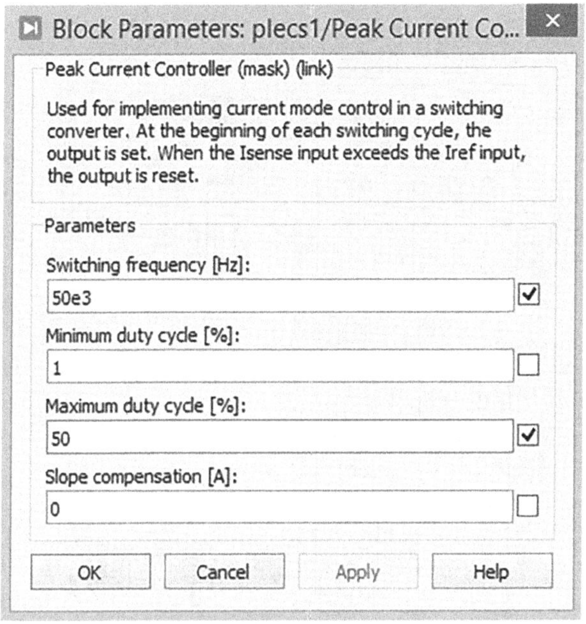

FIGURE 4.50 Setting the peak current controller parameters.

50%. We'll check the boxes, which will display the parameters on the schematic for us to see as depicted in Figure 4.50.

There we go, so next, what we want to do is we want to create a current sense signal to feed into our I_{sense} of the peak current controller. So, we're going to do that by simply placing a resistor component, which we'll grab from our passive electrical library and place that into our schematic, and we'll wire it up. We will change the resistor value to 0.5 ohms and we can display that. Now, we also need a voltmeter component. Again, utilizing the search function and we wire it up as illustrated in Figure 4.51.

So, here we go, and we've routed now our I_{sense} signal, and now for the reference current signal, I_{ref}, we're actually going to first use a constant block, so we're going to specify a set point value here, 0.1 amps as you can see in Figure 4.51. This is going to limit our actual peak inductor current to 10 amps and it's going to create an

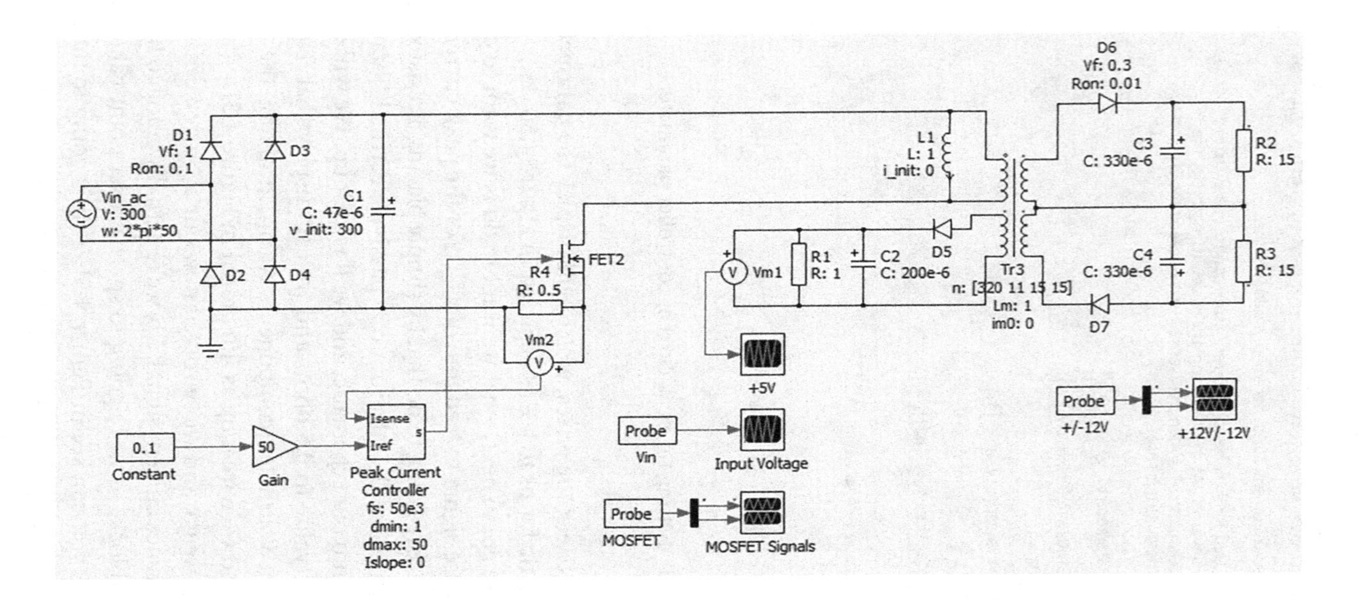

FIGURE 4.51 Wiring up the peak current controller signals.

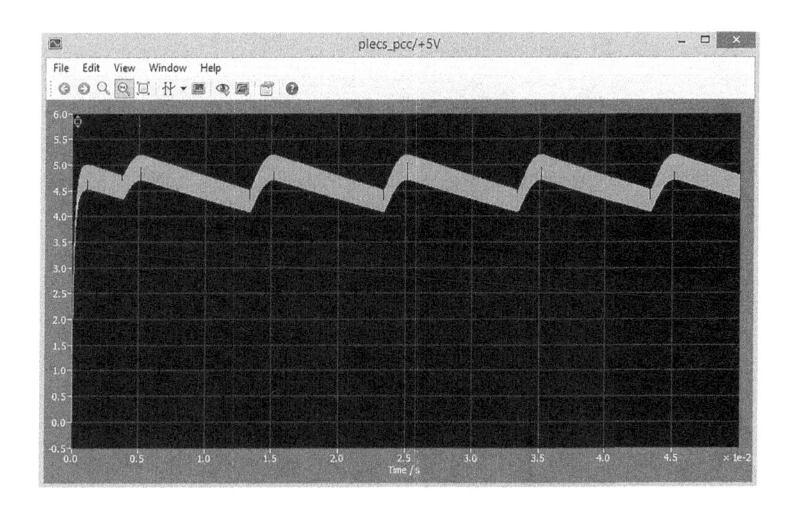

FIGURE 4.52 Regulation on 5 V with peak current controller.

output voltage of 5 V in other primary winding of flyback transformer. So, we can run this simulation again, and now you can see the regulation on the 5-V output as shown in Figure 4.52.

So, what's impractical about the controller that we just implemented? So, what's impractical about the converter system is that an outer, not typically in a system, we're going to have an outer voltage loop that provides this reference current, 0.1 A, for the current controller in order to regulate the output. To design the voltage controller, now we're going to use the small signal transfer function of the current-controlled converter to do that. We can also now get into these impulse response and AC sweep tools to get this function in the form of a Bode plot. So, to do that, we're going to open up another model as depicted in Figure 4.53.

So, we've pre-configured this model to perform this impulse response analysis. We will explain the changes that we've made. So, first things first, we have changed the input rectifier and capacitor, and we've changed it to a 300-V DC source component, the dynamics of the input rectifier are much slower than those of the flyback converter and can therefore be ignored. Simple simplifications like this don't affect our controller design, and they make

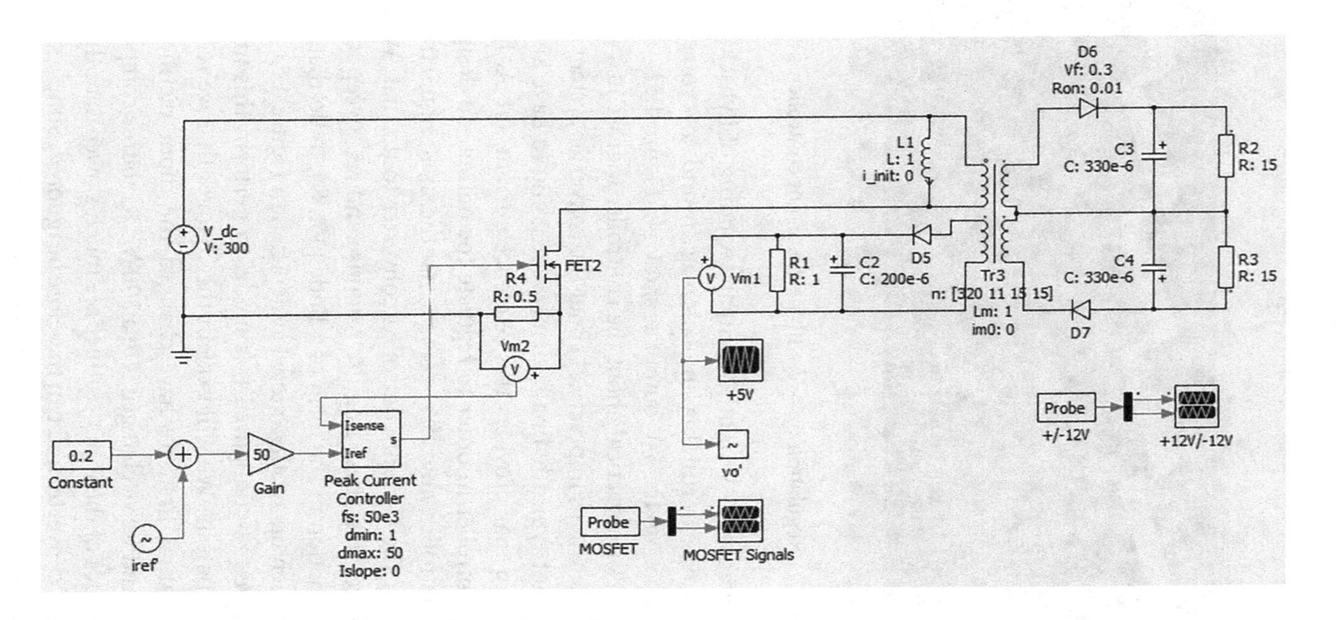

FIGURE 4.53 The impulse response analysis schematic.

this small signal analysis easier to execute. To obtain a Bode plot of the open voltage loop, we've added two special components to the model. So, you can see here, we've added what's called a small signal perturbation block (i_{ref}'). So, this component is added to our current set point value. We also need to add a small signal response to the perturbation to measure the perturbation at the 5-V output. So here's our small signal response block (vo′) that we're using to measure the output. These components are found in the controls category, and then found down there in the small signal analysis sub-category of the controls category. We've labeled these as i_{ref}' and vo′, respectively, and now we can open the configuration settings of the analysis by going to the simulation menu and going to analysis tools, and then click on the plus (+) button to add an impulse response analysis as shown in Figure 4.54.

So, there are a few settings to highlight as shown in Figure 4.55.

We've got the impulse response analysis selected as our choice of analysis. We've labeled it simply as impulse response analysis. This is a configuration you add and remove analyses as you see fit for your model. The system here, we've specified a period length, and it's the least common multiple of all the periodic sources in the model, so which in our case is the switching signal or 1 over 50 kHz. The

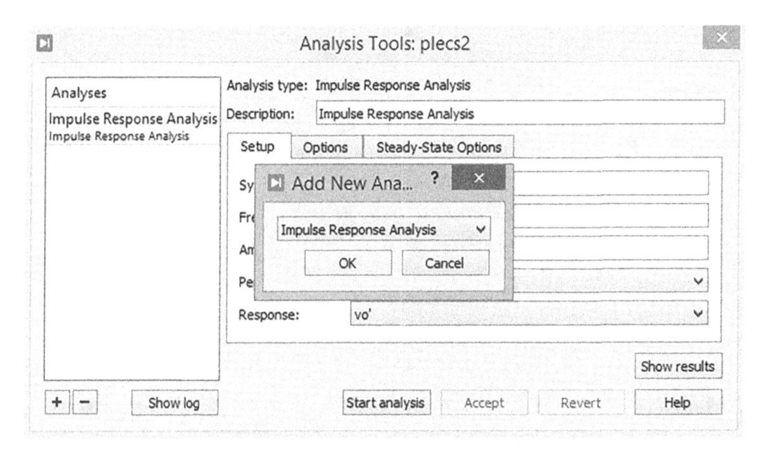

FIGURE 4.54 Adding an impulse response analysis.

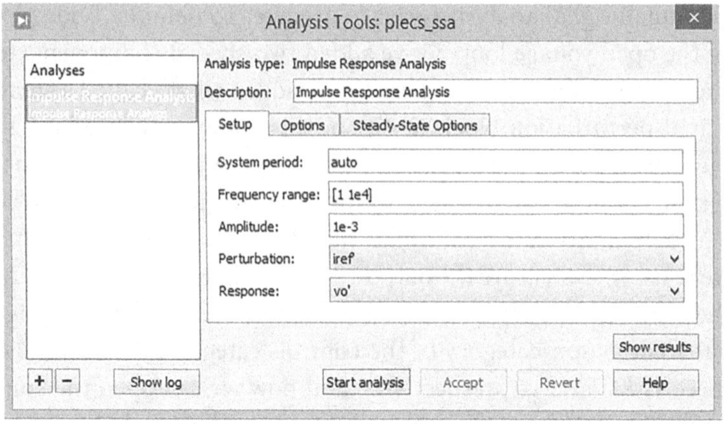

FIGURE 4.55 The settings for impulse response analysis.

frequency sweep range should be a vector that includes the lowest and the highest perturbation frequencies, and then on the perturbation and response section, we want to indicate the perturbation and response blocks that we have that we want to use for this analysis in our model. We only have one of each, so it's very easy for us as illustrated in Figure 4.55. In the options tab as shown in Figure 4.56, we can specify the number of automatically distributed frequencies.

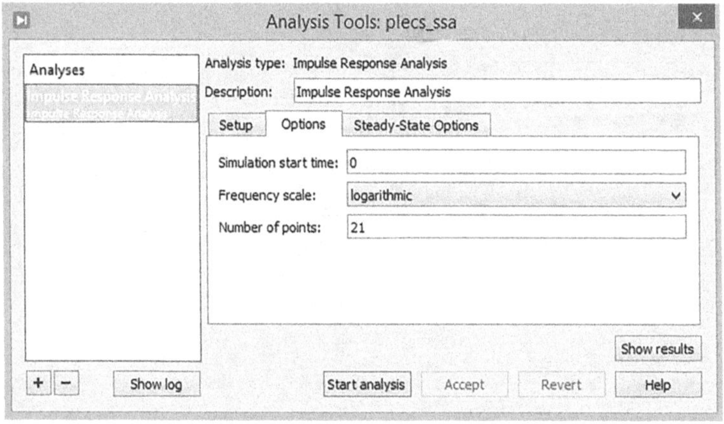

FIGURE 4.56 The options tab of analysis tools window.

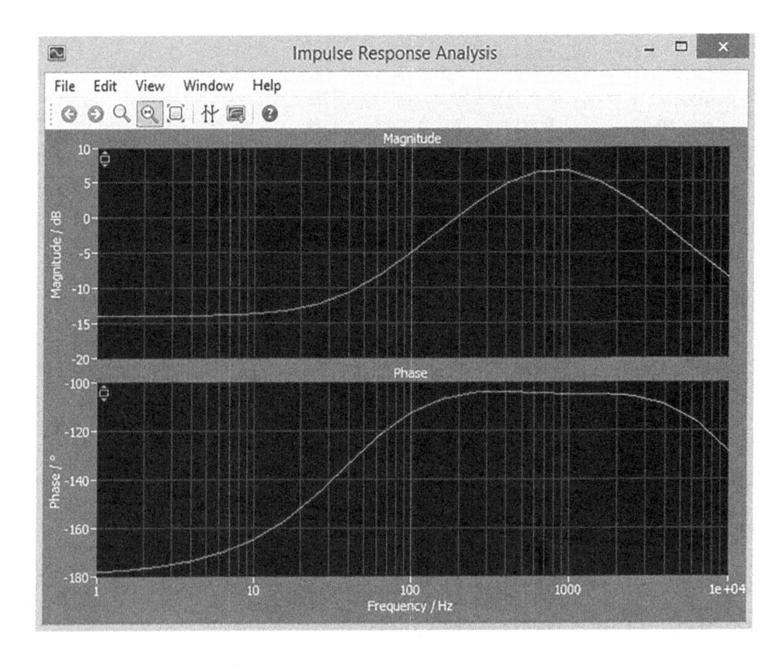

FIGURE 4.57 The Bode plot of impulse response analysis.

So, we're going to start this analysis, and then we're going to explain a few more things. So, that was very quick, we're done, that analysis was super-fast, and that's because the impulse response analysis is actually the fastest method that we have of generating a Bode plot as shown in Figure 4.57.

When we run the impulse response analysis, it actually first runs a steady-state analysis in order to find the stable operating point of the system. The steady-state analysis is based on a Newton Raphson iterative technique, and it may not converge when the system is run from startup because the system is under damped, and the state variables in PLECS are going to be too far from the final operating point. So, the simplest method of addressing this problem is to simulate the system for a specified number of initialization cycles before the steady-state analysis is run. If we go back to our configuration settings and we go to the steady-state options,

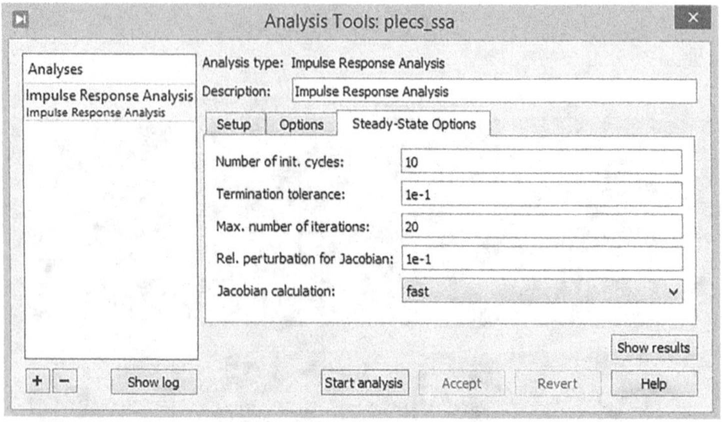

FIGURE 4.58 The steady-state options tab of analysis tools window.

we can see here in Figure 4.58 that we have the ability to specify this number of initialization cycles.

This then causes the Jacobian matrix to be calculated from an operating point that is closer to the steady-state operating point, which increases the likelihood of convergence in the under-damp system and thus allowing the analysis to run to completion. So, that's our analysis, so we can go back to this frequency response. We can see the frequency response here of our current controlled flyback converter. We're now going to save this data. So, we'll go file export as a CSV file as depicted in Figure 4.59.

So, we'll save it as flyback.csv, so now at this stage, we are ready to start talking about that voltage controller. So, what we're going to use MATLAB for a PID controller design. So, in the MATLAB command window, we write the following codes.

```
data = csvread('C:\CRC Press\flyback.csv');
w = 2*pi*data(:,1);
val=10.^(data(:,2)/20).*exp(j*data(:,3)*pi/180);
```

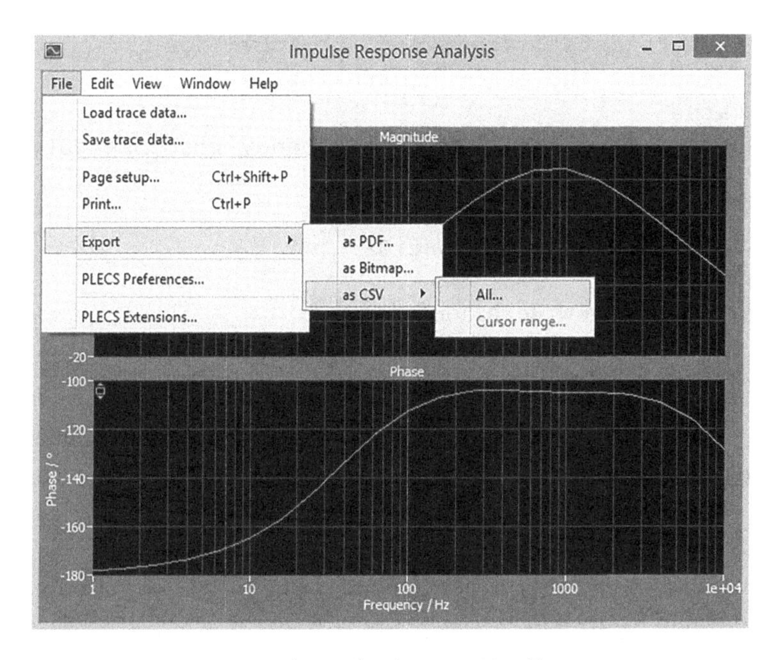

FIGURE 4.59 Exporting the Bode plot as a CSV file.

```
sys = frd(val,w);
bode(sys),grid minor
G = tfest(sys,3)
```

The estimated transfer function using MATAB with an accuracy of 99.71% is as below.

```
G =
   -4.024e04 s^2 + 5.83e09 s + 1.338e12
  ---------------------------------------------
  s^3 + 2.596e05 s^2 - 4.657e08 s - 6.749e12
 Continuous-time identified transfer function.
Parameterization:
  Number of poles: 3   Number of zeros: 2
  Number of free coefficients: 6
```

```
  Use "tfdata", "getpvec", "getcov" for
parameters and their uncertainties.
Status:
Estimated using TFEST on frequency response data
"sys".
Fit to estimation data: 99.71%
FPE: 1.067e-05, MSE: 8e-06
```

Then, we can write the following code.

```
pidTuner(G)
```

We can adjust the response time and transient behavior slides to get the step response as illustrated in Figure 4.60.

Then, we click on the Show Parameters button to see the PID controller parameters alternatively, we can click on the Export button and give it a name (Cg) and click on the OK button as shown in Figure 4.61.

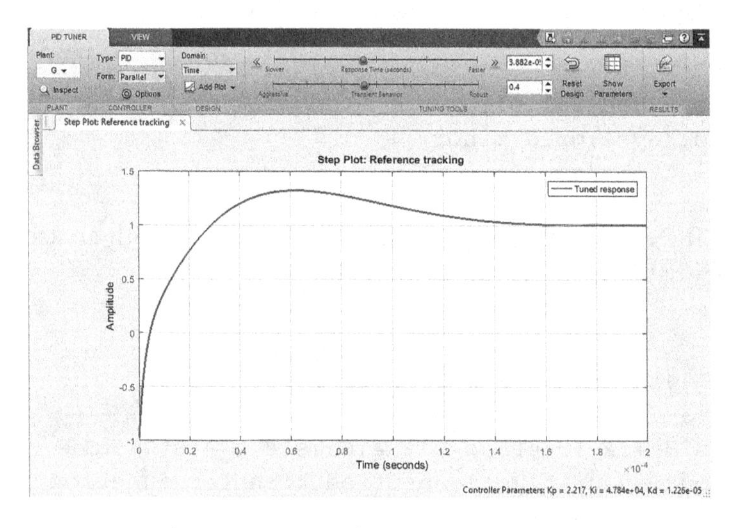

FIGURE 4.60 The PID tuner window.

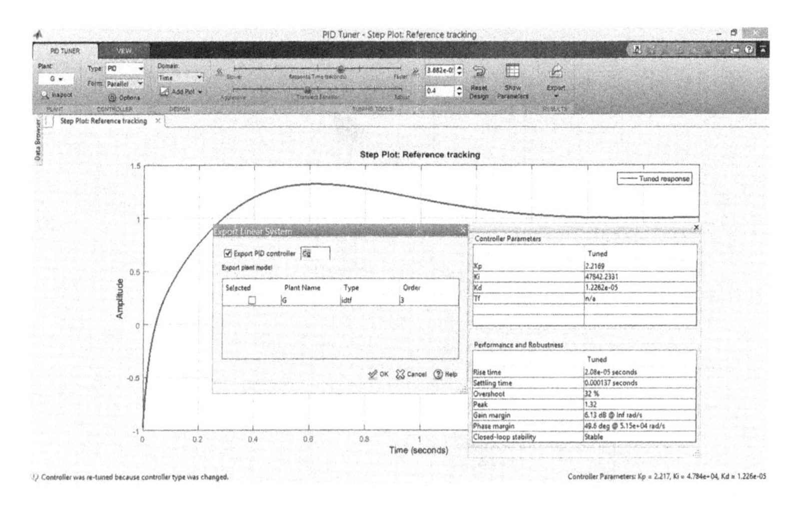

FIGURE 4.61 The show parameters and export windows.

Next, in the command window of MATLAB, we can type the 'Cg' and press the enter key to get the PID controller parameters as below.

```
>> Cg
Cg =
                  1
  Kp + Ki * --- + Kd * s
                  s
  with Kp = 2.22, Ki = 4.78e+04, Kd = 1.23e-05
 Continuous-time PID controller in parallel
form.
```

Now, we can add the Continuous PID Controller block in Figure 4.53 to create a closed-loop voltage and peak current controller as depicted in Figure 4.62.

The waveform of 5 V scope is illustrated in Figure 4.63.

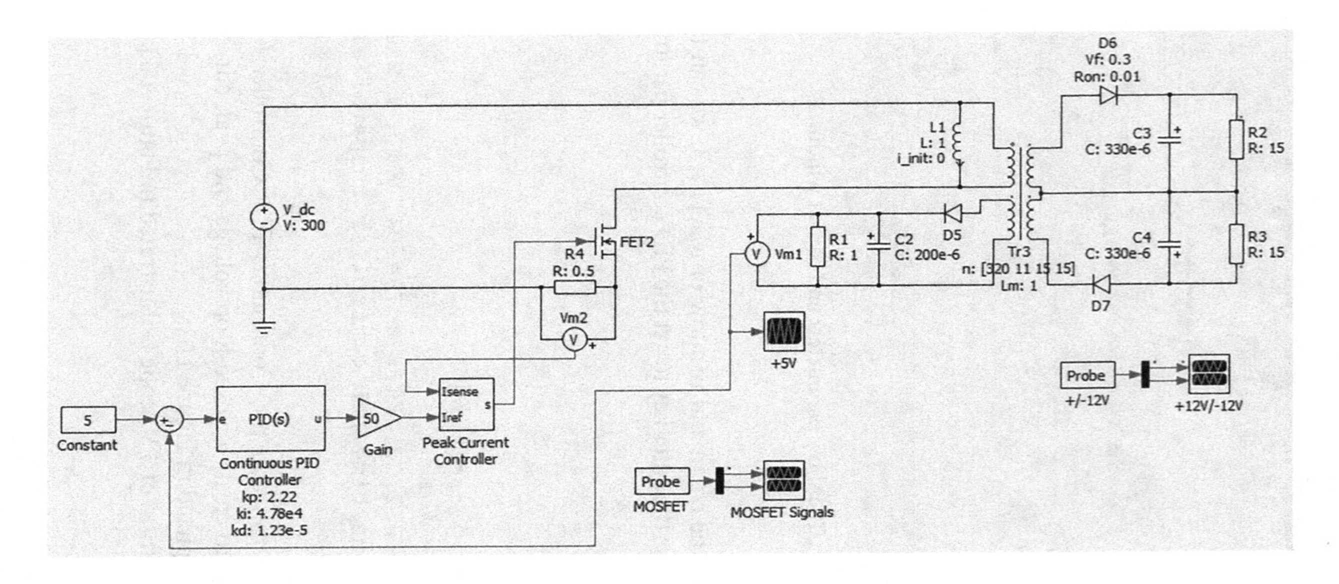

FIGURE 4.62 The closed-loop voltage and peak current controller.

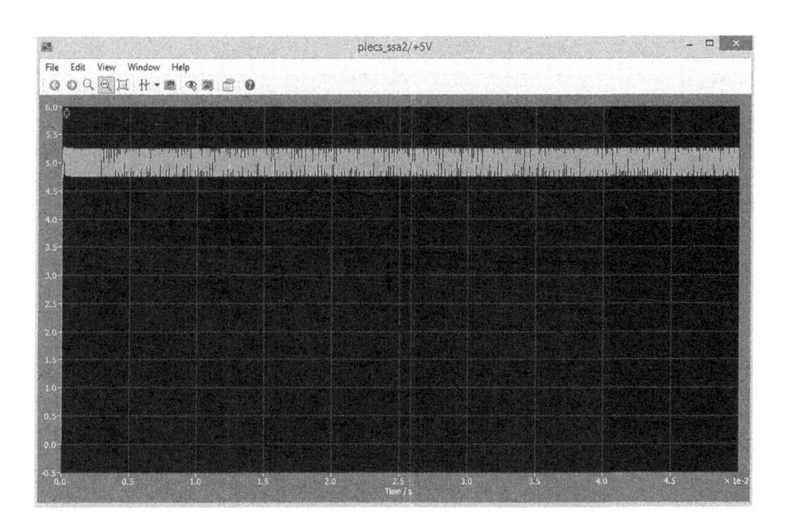

FIGURE 4.63 The waveform of 5 V scope.

Then, we add a load step to the model, to evaluate the transient behavior of closed-loop controller. So, at the time 25 ms, we add a parallel resistor (2 Ω) in the 5 V circuit as shown in Figure 4.64.

The output voltage waveforms during the load step at the time 25 ms are shown in Figure 4.65.

Therefore, our closed-loop controller is stable. In order to code generation for the closed-loop voltage and peak current controller, we use a subsystem as shown in Figure 4.66 and take our controller inside of it. Then we use STM32 target elements such as ADC and digital out as depicted in Figure 4.67.

If we run the schematic of Figure 4.66, the output voltage waveforms remain the same as Figure 4.65 as shown in Figure 4.68.

Then, we do right click on the subsystem and select Subsystem and Execution settings... and in the popped-up window we tick the Enable code generation and choose the Discretization step size as 2e-7 as illustrated in Figures 4.69 and 4.70, respectively.

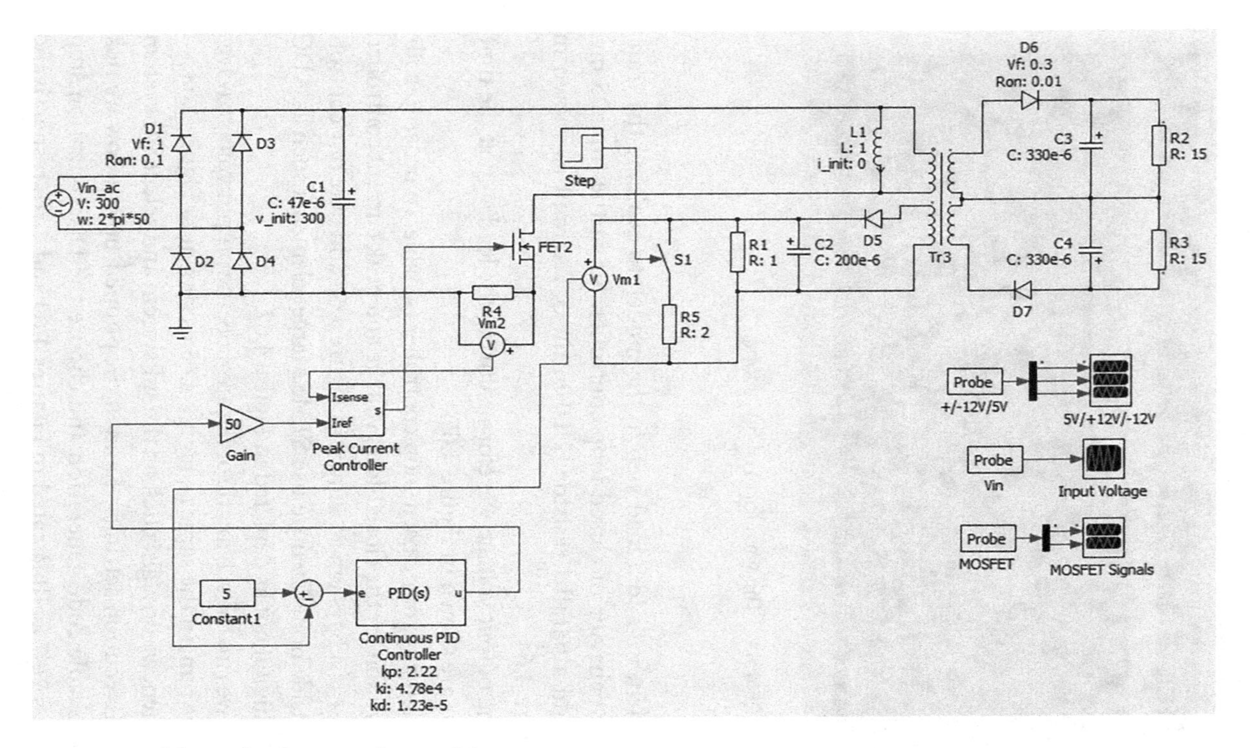

FIGURE 4.64 Adding a load step to the model.

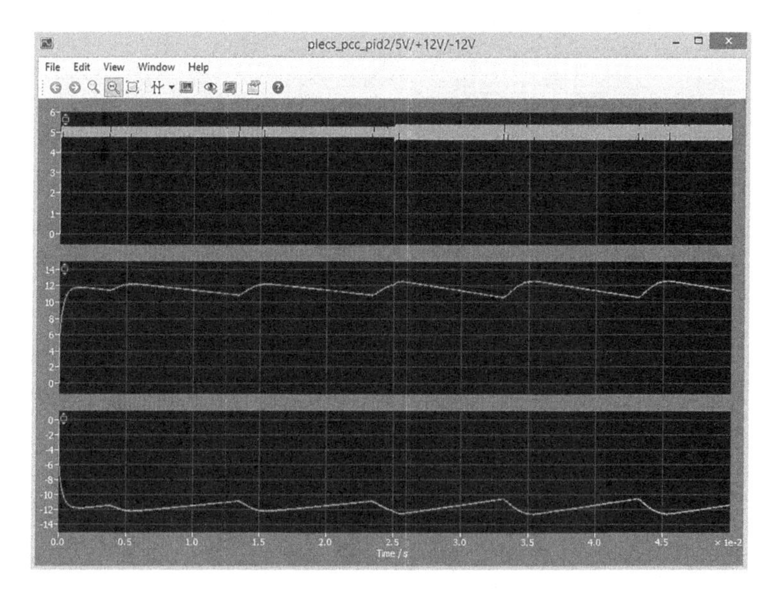

FIGURE 4.65 The output voltage waveforms during the load step.

In the Open projects from File System window, we click on the Archive... button as illustrated in Figure 4.29, to open the g474.zip file in the following directory as shown in Figure 4.30.

```
PLECS4.7(64 bit) > tsp_stm32 > projects > g474.
zip
```

Then we run the STM32CubeIDE to enter it and from File icon we select Open projects from File System... option as depicted in Figure 4.28. Then we do click on the Finish button in the Import Projects from File System or Archive window as shown in Figure 4.31. In the Project Explorer, we find the cg folder and then right click on it and select the Properties option as depicted in Figure 4.32. Therefore, the Properties for cg window is opened and we do copy the Location directory as illustrated in Figure 4.71.

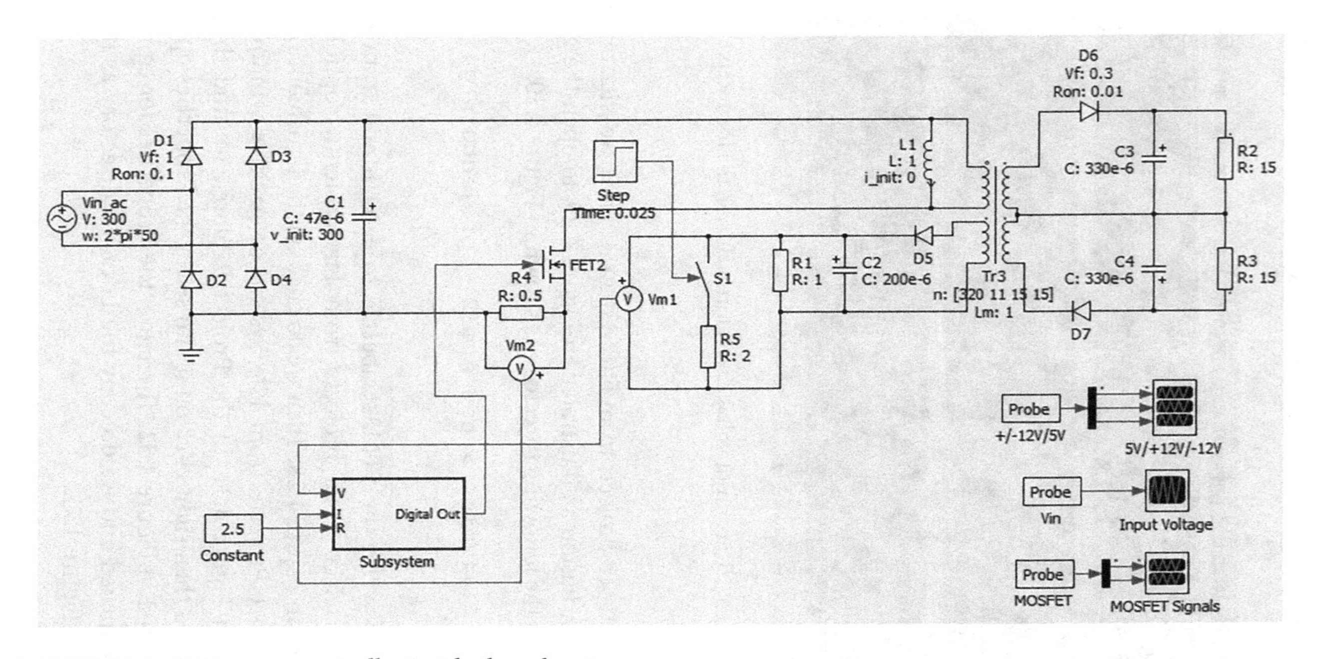

FIGURE 4.66 Taking our controller inside the subsystem.

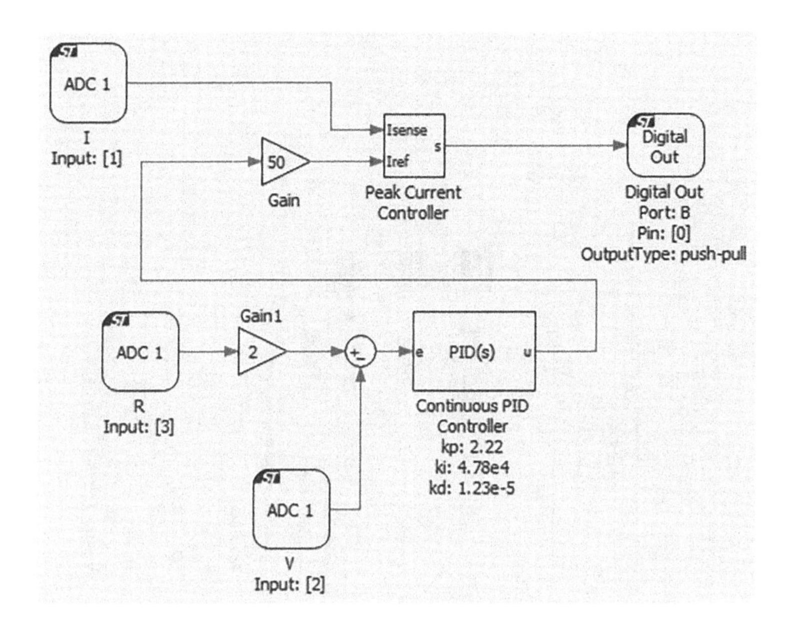

FIGURE 4.67 The subsystem controller and STM32 target elements.

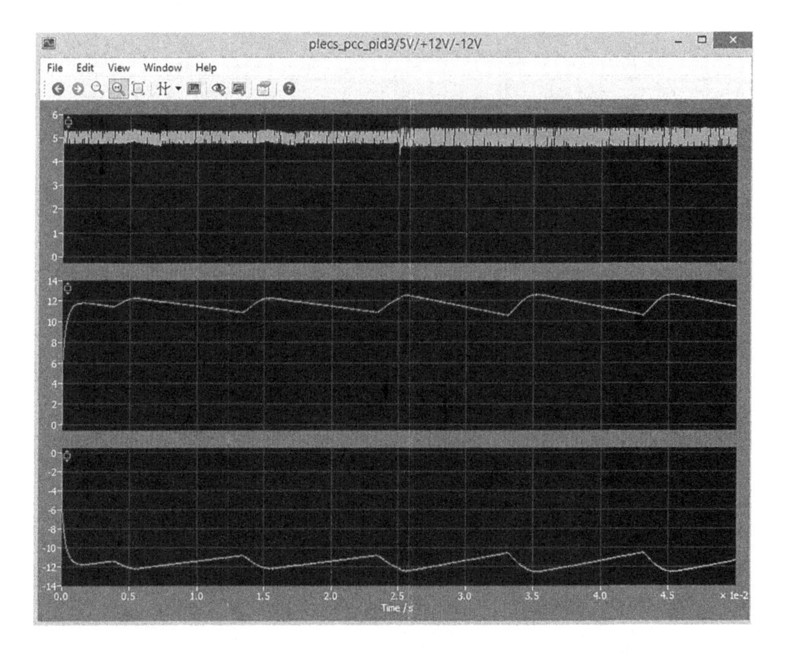

FIGURE 4.68 The output voltage waveforms with STM32 target elements.

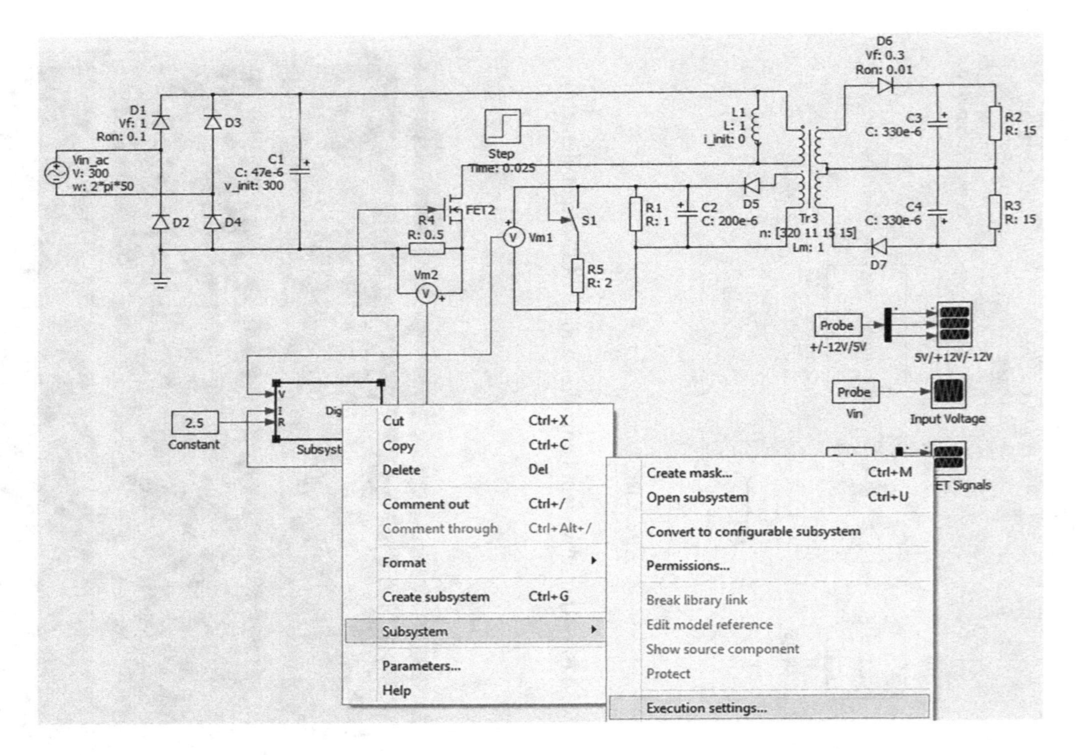

FIGURE 4.69 Selecting the execution settings… option.

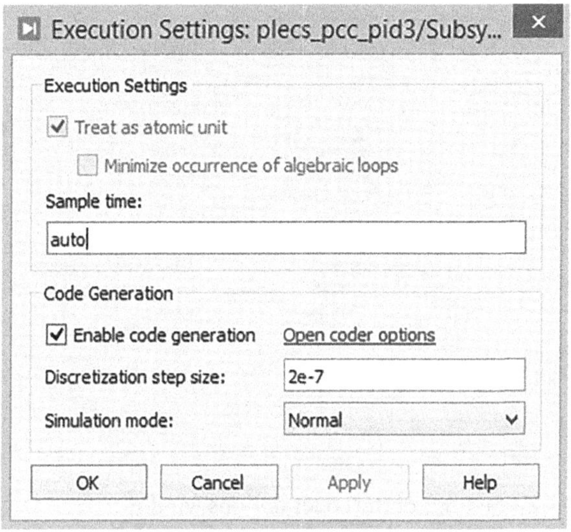

Execution Settings: plecs_pcc_pid3/Subsy...

Execution Settings

Treat as atomic unit

Minimize occurrence of algebraic loops

Sample time:

auto

Code Generation

Enable code generation Open coder options

Discretization step size: 2e-7

Simulation mode: Normal

OK Cancel Apply Help

FIGURE 4.70 Enabling code generation and setting discretization step size.

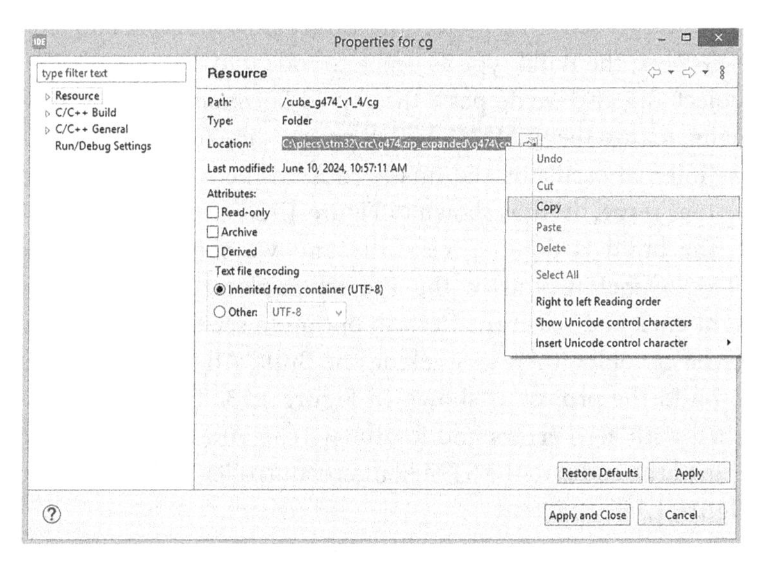

FIGURE 4.71 Doing copy the location directory.

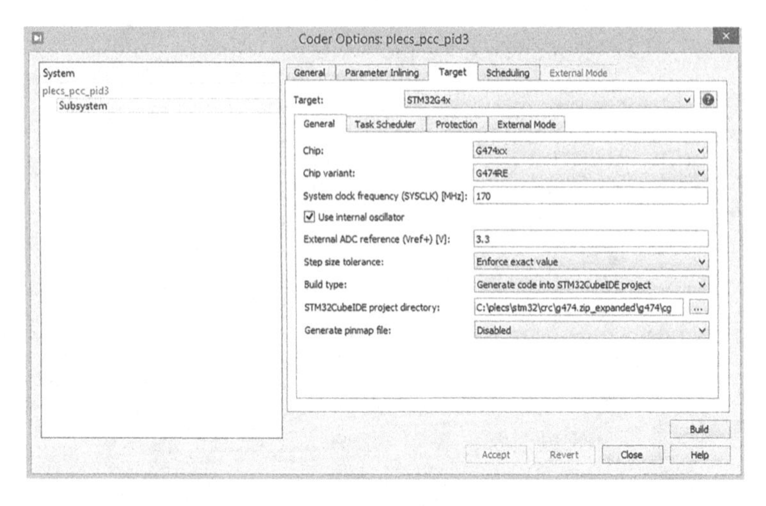

FIGURE 4.72 Settings of the coder options window.

Then, we close the Properties for cg window and come back to the PLECS environment and in the Coder icon of menu we select the Coder options…, to open the Coder Options window and in the Target section we select the Target as STM32G4x and chip as G474xx, the Build type as generate code into STM32CubeIDE project and also we do paste the copied Location directory as in Figure 4.71 to the STM32CubeIDE project directory and also tick use internal oscillator and finally click on the Accept and Build buttons respectively as shown in Figure 4.72.

The build is done successfully and we come back to the STM32CubeIDE and in the Project Explorer section we do right click and select the Refresh option to see the codes added to the cg folder, then we click on the Build All (Ctrl+B) button to build the project as shown in Figure 4.73. The build is finished with zero errors and finally, we can run the program to upload the codes to the STM32 microcontroller as illustrated in Figure 4.74.

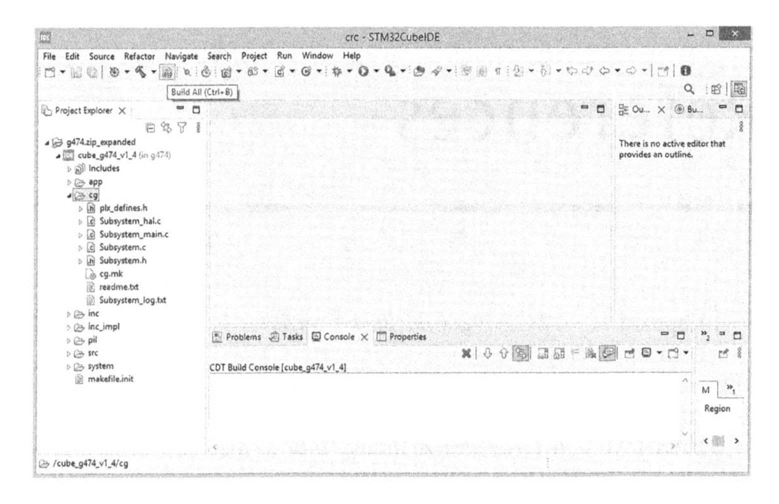

FIGURE 4.73 Clicking on the build all (Ctrl+B) button.

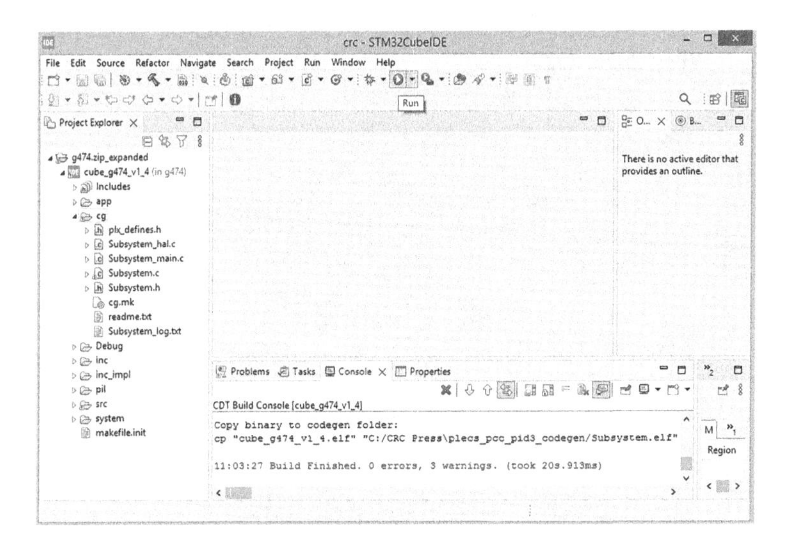

FIGURE 4.74 Successfully building and running the program.

References

[1] Kupolati H. *STM32 Practical Buck Converter Implementation and Control*, Udemy, 2023.

[2] Nabil M. *Digital Feedback Control Tutorial with Arduino*, Udemy, 2019.

[3] Pakdel M. *Advanced Programming with STM32 Microcontrollers*, Elektor, 2020, 216 p.

[4] DC Motor Speed: System Modeling. https://ctms.engin.umich.edu/CTMS/index.php?example=MotorSpeed§ion=SystemModeling

[5] Pakdel M. *Advanced Modeling and Control of DC-DC Converters*, Wiley, 2025, 352 p.

Index

Note: Pages in *italics* refer to figures and pages in **bold** refer to tables.

E

F

G

H

I

K

M

N

O

P

S

Printed in the United States
by Baker & Taylor Publisher Services